STOCHASTIC MODELS OF CARCINOGENESIS

STATISTICS: Textbooks and Monographs

A Series Edited by

D. B. Owen, Coordinating Editor
Department of Statistics
Southern Methodist University
Dallas, Texas

R. G. Cornell, Associate Editor
for Biostatistics
University of Michigan

W. J. Kennedy, Associate Editor
for Statistical Computing
Iowa State University

A. M. Kshirsagar, Associate Editor
for Multivariate Analysis and
Experimental Design
University of Michigan

E. G. Schilling, Associate Editor
for Statistical Quality Control
Rochester Institute of Technology

ADDITIONAL VOLUMES IN PREPARATION

STOCHASTIC MODELS OF CARCINOGENESIS

WAI-YUAN TAN

Memphis State University
Memphis, Tennessee

Marcel Dekker, Inc. New York • Basel • Hong Kong

Library of Congress Cataloging--in--Publication Data

Tan, W. Y.
    Stochastic models of carcinogenesis/W. Y. Tan
        p. cm. -- -- (Statistics, textbooks and monographs; vol. 116)
    Includes bibliographical references and index.
    ISBN 0-8247-8427-8 (alk. paper)
    1. Carcinogenesis-- --Mathematical models. 2. Stochastic analysis.
I. Title. II. Series
    [DNLM: 1. Models, Biological. 2. Models, Statistical.
3. Neoplasms-- --epidemiology. 4. Neoplasms-- --etiology. 5. Stochastic
Processes.   QZ 202 T16s]
RC268.5. T36 1991
616.99'4071-- --dc20
DNLM/DLC
for Library of Congress   91-6835
                                                                   CIP

This book is printed on acid-free paper.

MARCEL DEKKER, INC.
270 Madison Avenue, New York, New York 10016

Current printing (last digit):
10 9 8 7 6 5 4 3 2 1

PRINTED IN THE UNITED STATES OF AMERICA

To my wife
SHIOW-JEN TAN

Preface

In the past 15 years, major advances and discoveries have been made by molecular biologists and geneticists in cancer genes and their functions and in the mechanism of carcinogenesis. For research on the control and prevention of cancer, it is essential that the biological findings be transformed into mathematical models. With that in mind, the purpose of this book is to present an up-to-date survey of mathematical models of carcinogenesis.

This book is organized into six chapters. In Chapter 1, a brief review of the most recent developments in cancer biology is presented. In Chapter 2, some basic mathematical concepts and stochastic processes that are useful for the development of stochastic models of carcinogenesis are given. Some stochastic models of carcinogenesis are then studied in Chapters 3 through 6. These models are motivated by recent findings of cancer biology as reviewed in Chapter 1. Chapter 1 may thus help familiarize mathematicians and statisticians with the most recent developments of cancer biology, while Chapter 2 may help medical doctors and biologists understand some basic stochastic processes that are useful in the development of stochastic models of carcinogenesis. Chapter 3 is principally a bringing together of research conducted over the years by Professor S. H. Moolgavkar of the University of Washington at Seattle, Washington, together with his colleagues, and by myself together with my colleagues, Dr. C. C. Brown of NCI (National Cancer Institute), Dr. K. Singh of the University of Alabama, Birmingham, Alabama, and Dr. M. A. Gastardo. The contents of Chapters 4, 5, and 6 are basically taken from research I conducted over the years with my colleagues, Dr. C. C. Brown and Dr. K. Chu of NCI and Dr. K. Singh of the University of Alabama, Birmingham, Alabama.

I originally compiled this book in 1987 for a course for my students in the Department of Mathematical Sciences at Memphis State University, Memphis, Tennessee, when I was giving a series of lectures on stochastic models of carcinogenesis. It was intended for applied mathematicians, applied probabilists, biostatisticians, cancer researchers, as well as environmental scientists who are interested in quantitative risk assessment of environmental agents. As a unique feature of the book, I have provided the most recent findings of cancer biology as evidence for the models. To make the book a useful reference for applied mathematicians and statisticians, as well as for cancer researchers and theoretical biologists, I have given an extensive bibliography of cancer biology and in-depth mathematical analysis for each of the models. Thus, the book may be used as a reference for courses on stochastic processes and a textbook for seminar courses in biomedical research and cancer research.

I would like to thank Dr. C. C. Brown, Dr. M. H. Gail, and Dr. D. Byer of NCI for their friendship, encouragement and support. I want also to express my sincere appreciation to my wife, Shiow-Jen, my daughter, Emy, and my son, Eden, for their support and patience with my devotion to research and study.

Finally, I wish to thank Mrs. Shannon B. Diamond and Mrs. Helen Wheeler for their excellent typing of my manuscript and Ms. Maria Allegra of Marcel Dekker, Inc., for her assistance in the publication of my book.

Wai-Yuan Tan

Contents

Contents

STOCHASTIC MODELS OF CARCINOGENESIS

1

Cancer Biology and the Multistep Nature of Carcinogenesis

In this chapter, a brief review of cancer biology is presented. In Section (1.3), the multistep nature of carcinogenesis is demonstrated through initiation and promotion studies and through the cell culture studies of Barrett and co-workers on the development of cancer tumors of rat tracheal cells and Syrian hamster embryo fibroblasts. Since the genetic changes are the dynamics of carcinogenesis, in Section (1.4) we therefore give some specific examples of genetic changes which have been identified with some processes of carcinogenesis. For completeness and for understanding the disease of cancer, in Section (1.5) we present the cancer genes which include oncogenes, antioncogenes and accessory cancer genes, together with some of their basic properties. Then, in Section (1.6), we illustrate how oncogenes and antioncogenes work at the molecular level. By scrutinizing and integrating the recent biological findings of cancer, we formulate in Section (1.8) some models of carcinogenesis; these models will be studied in detail in Chapters 3–6.

Finally, to complete this chapter, in Section (1.9) some comments and criticisms are directed to the classical Armitage–Doll multistage model of carcinogenesis.

1.1 INTRODUCTION

Cancer is a deadly disease which remains uncontrollable in most cases. In order to develop an effective cure and control of cancer, basic understanding of the mechanisms leading to the cancer phenotype from normal cells is essential and highly desirable.

To seek answers to the cancer problems, two groups of scientists around the world have been approaching the problems from different angles with different emphases and different levels of success. One group of scientists involves molecular biologists and geneticists who have done extensive laboratory work to uncover the basic steps and fundamental aspects of carcinogenesis. Another group of scientists involves epidemiologists and mathematicians who have been trying to develop mathematical models for carcinogenesis to interpret human epidemiological data and animal experimental data. In the past ten years major advances and important discoveries have been made by molecular biologists and geneticists; compared to the biological findings, however, epidemiologists and mathematicians appear to be falling far beyond in their studies of carcinogenesis.

In order for the biological findings to be tested quantitatively against human epidemiological data and animal experimental data, and in order to develop efficient controlling strategies for cancer, it is essential that the biological findings be transformed into mathematical models. To further studies of mathematical models of carcinogenesis, the purpose of this monograph is therefore to present an up to date survey of mathematical models of carcinogenesis. An excellent survey of this nature has been given by Whittemore (157) and by Whittemore and Keller (158) in 1978. However, biological findings in the past ten years have invalidated some of the carcinogenesis models (e.g. the Armitage–Doll multistage model, see Section 1.9); furthermore, some new useful carcinogenesis models have been proposed since 1978. The present survey may therefore be viewed as a continuation of Whittemore and Keller (158) and Whittemore (157). For providing some biological support for the carcinogenesis models, we give in this chapter a brief review of the most recent biological findings in cancer and illustrate how the cancer genes work. This is absolutely essential for the development of reasonable and biologically meaningful stochastic models of carcinogenesis.

1.2 THE STEM CELL THEORY OF CARCINOGENESIS

Biologists have discovered that all tissues consist of two types of cells, the differentiated cells which are major components of the tissue proper and the stem cells from which cancer tumors may develop. This stem cell theory stipulates that only stem cells can divide giving rise to new stem cells and differentiated cells; the differentiated cells do not divide and are end cells to perform specific functions of the tissue. The above stem cell theory also applies to cancer tumors (21, 94); that is, cancer tumors also consist of differentiated tumor cells and stem tumor cells. The following results appear to provide strong evidence for the stem cell theory of cancer tumors.

i. Clinical experience with radiation therapy suggests that only a small fraction of tumor cells would have to be killed to abolish tumor regeneration (24).
ii. Fractionation of suspensions of cells from human tumors has demonstrated that proliferative activity, self-renewal capacity and cell differentiation features are restricted to separate subpopulations with defined physical properties (16, 93–95).

1.3 MULTISTEP NATURE OF CARCINOGENESIS

It is now universally recognized that cancer tumor develops from a single normal stem cell which has undergone a series of genetic changes (4, 9–12, 28, 46–49, 52, 113, 131, 151, 162). Geneticists and molecular biologists have identified genetic changes for carcinogenesis (8–10, 18–20, 22–23, 59, 62, 76–80, 112, 119, 131, 141, 148–151). Borek and Sachs (17), Kalunaga (71), Farber (47) and Farber and Sarma (49) showed that at least one round of cell proliferation was required for a genetic change to become permanent. The initiation—promotion—initiation experiments on rat skin by Hennings et al (64) and on rat liver by Scherer et al (120) have demonstrated that carcinogenesis starts by a genetic change; the initiated cells then undergo cell proliferation; then another genetic change takes place, etc. These and many other biological findings imply that carcinogenesis involves a series of heritable changes related to the cell's release from growth control sequentially (4, 9–11, 46–49, 101,

148, 150, 151); further, cells in each step are subjected to considerable random variation related to cell proliferation and differentiation (21, 28, 71, 72, 94, 99, 100). That is, carcinogenesis is a multistep and sequential random process with each step involving interplay of genetic factors, epigenetic factors within stem cells and interactions between cells (cell communication, see 114, 142–145).

1.3.1 An Experimental Demonstration of the Multistep Nature of Carcinogenesis

The multistep nature of carcinogenesis has been demonstrated experimentally by Barrett and his coworkers by the cell culture method on rat tracheal epithelial cells (RTE cells, see 107) and on Syrian hamster embryo fibroblasts (SHE, see 11).

When the RTE cells were treated by a carcinogen such as MNNG (N-methyl-N'-nitro-N-nitrosoguanidine) and then promoted by TPA (12-0-tetradecanoyl-phorbol-13-acetate), about 2–5 percent of RTE cells developed into EG variants (enhanced growth cells). The RTE normal cells and EG variants have finite life-spans (i.e. can divide only a finite number of times) but EG variants continue to proliferate after normal RTE cells stop dividing. The EG variants can further be induced by carcinogens to develop immortal EG variants which have an infinite life-span (i.e. can divide indefinitely in culture). The immortal EG cells are anchorage dependent (ag^-, i.e. cannot grow in soft agar) but can be induced by carcinogens to develop anchorage-independent (ag^+, i.e. can grow in soft agar) immortal EG variants. These ag^+ immortal EG variants are nontumorigenic (tum^-) when injected into normal rat but can further be induced by carcinogens to become tumorigenic (tum^+) neoplastic variants. (see Figure 1.1).

Hybridization analyses show that both immortalization and anchorage independence are genetic changes which can arise either spontaneously with low frequency or can be induced by carcinogens or by transfection of oncogenes, with nuclear oncogenes such as myc, myb, adeno-virus E1A, polyoma large T, SV-40 large T and P53 related to immortalization while with cytoplasmic oncogenes such as ras or src related to anchorage independence (see Section 1.5). Cell culture studies of the SHE fibroblasts show that normal SHE cells, immortalized cells (preneoplastic) and neoplastic cells (transformed cells) are distinctly different from one another (11). Normal SHE cells are diploid and morphologically normal cells, have finite life-span, cannot be transformed by ras or src oncogene alone,

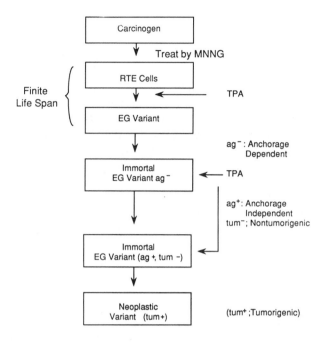

ag⁻ ⟶ ag⁺   (Appear after 5 – 10 passages);
ag⁺, tum⁻ ⟶ tum⁺   (Appear during 10 – 20 passages).

Figure 1.1   Multistep *in vitro* transformation of rat tracheal epithelial (RTE) cells. (From Nettesheim and Barret 1985.)

and are nontumorigenic when attached to a plastic substrate. On the other hand, immortalized SHE cells (preneoplastic) are aneuploid and morphologically transformed cells, have infinite life-span, can be transformed by ras or src oncogene alone and are tumorigenic when attached to a plastic substrate; immortalized cells, however, are anchorage dependent (cannot grow in soft agar) and are nontumorigenic in suspension when injected into normal cells. The neoplastic cells, besides being aneuploid, morphologically transformed, immortal and tumorigenic when attached to plastic substrate, are also anchorage-independent (can grow in soft agar) and tumorigenic in suspension (11).

The above demonstration implies that under normal conditions, both immortalization and transformation are necessary steps for the process of carcinogenesis. In fact, both the nontumorigenic immortalized cell lines and the nontumorigenic transformed cell lines with finite life-span have been isolated. For example, Hurlin, Maher and McCormick (70) isolated a diploid human fibroblast cell strain with an infinite life-span (immortalized), called MSU-1.1, by transfecting foreskin-derived normal diploid human fibroblasts with a plasmid carrying a v-myc oncogene. The MSU-1.1 lines show normal morphology, have the same diploid karyotype as normal cells, do not form foci in soft agar (anchorage-dependent) and are nontumorigenic. Also, by transfecting diploid human fibroblast cells with a plasmid which contains the human T24 H-ras oncogene flanked by Moloney murine sarcoma virus long terminal repeats and simian virus 40 transcriptional enhancer sequences, Hurlin et al (69) obtained a transformed cell line. This line develops distinct foci and forms colonies in soft agar (anchorage independent) yet does not acquire an infinite life span and is not tumorigenic.

The above illustrations imply that under normal environmental conditions at least two steps (immortalization and transformation) are required for the process of carcinogenesis. The following biological findings provide evidence indicating that in some cases immortalization and transformation are not sufficient for the tumorigenic conversion of normal stem cells:

i. Working with RTE cells, Thomassen et al (140) observed different responses to the induction (both spontaneous and induced) of anchorage independence from different immortal cell lines; the rate of acquisition of anchorage independent growth from immortalized cell lines varies from 10^{-4} per cell per generation to 10^{-7} per cell per generation. These results indicate the heterogeneous nature of the immortalized cell lines.

ii. Ruley et al (118), Tsunokawa et al (146) and Thomasson et al (140) observed the existence of immortalized cell lines which cannot be transformed by oncogene ras or src alone.

iii. Hybridization studies between normal stem cells and immortalized cells (129) and between normal SHE cells and tumorigenic cells (9, 10, 12) indicate that both immortalization and anchorage-independence are genetic changes which behave like recessive traits; yet oncogene actions appear to be dominant (14, 22, 96, 131, 148, 151). This and other evidence (9, 12) suggest the existence of suppressor genes (i.e. antioncogenes, see Section 1.5.2).

iv. Oshimura, Gilmer and Barrett (111) and Barrett et al (12) observed in SHE cells that transfections by myc plus *ras* oncogenes are not sufficient to induce tumorigineity of SHE cells; a non-random loss of chromosome 15 has been observed in all tumorigenic cells besides the activation of the oncogenes *ras* and *myc*.

v. Land et al (85) showed that while *ras* and *myc* oncogenes cooperated to induce tumors in nude mice, the tumors grew only to a maximum size of 2 cm. By introducing a retroviral vector containing both v-ras and v-myc into mid-gestation mouse embryos, Compere et al (33) found that the *ras* and *myc* oncogenes cooperated to induce tumors in a wide variety of organs including brain, skin and kidney but the rates of tumor induction were quite different in different organs; also, while the majority of tumors in brain neoplasm, skin squamous cell carcinoma, kidney neoplasm and spindle cell neoplasms in heart, skin and subcutaneous tissue, are malignant, there are also benign tumors in some skin tumors (e.g. surface epithelial hyperplasia with severe dysplasia and mixed appendage tumor in skin).

vi. Sinn et al (122) obtained two separate transgenic mice strains, one carrying the MMTV/v-Ha-ras oncogene while the other carrying the MMTV/c-myc oncogene. These strains transmit their oncogenes to their progenies in a Mendelian fashion and show some benign proliferation of the lacrimal epithelium in the harderian gland. By interbreeding these two strains, Sinn et al (122) found that many of the F_1 individuals developed more rapidly malignant tumors of mammary, salivary and lymphoid tissues but the tumors were developed stochastically and clonally. These results imply that while both the *ras* and the *myc* oncogenes are required for developing malignant tumors, at least one more event is needed for the development of tumors.

Recently, it was shown in tissue cultures that the immortalization of cells and the ability of conferring responsiveness to the transforming effects of oncogenes like *ras* were separate properties of the oncogenes like *myc* or E1A (54, 151). Another convincing piece of evidence for the multistep nature of carcinogenesis came from recent studies on colon cancer and lung cancer in human beings. For colon cancer at least five cancer genes have been involved, including the oncogenes *ras* and *myc* and the antioncogenes from chromosomes 5q, 17p and 18q. For the squamous-cell carcinoma, large-cell carcinoma and adenocarcinoma of the human lung, at least nine cancer genes have been involved, including the oncogenes *myc, H-ras, K-ras, raf* and *jun* and the antioncogenes from chromosomes 3p, 11p, 13q and 17p (155).

1.3.2 Initiation, Promotion and Progression

Animal experiments (mouse or rat) involving different tumor sites (e.g. skin, liver, breast and bladder) have identified and characterized three phases of the carcinogenesis process, initiation, promotion and progression (9–11, 46–49). The classical painting experiments of mouse skin is to expose animals to a chemical called initiator such as DMBA (7,12-dimethyl-benz(a) anthracene) and then treat the animals with a chemical called promoter such as TPA (12–0-tetradecanoylphorbol-13-acetate). This type of experiment would induce development of papillomas from normal mouse skin. Hennings et al (64) reported that papillomas might further be induced to form carcinomas by initiators such as DMBA but not by promoters. Since initiators induce genetic changes while promoters only facilitate proliferation of initiated cells (9, 10, 11, 125, 126), the experimental results of Hennings et al (64) indicated that carcinogenesis started with a genetic change which is induced by an initiator; cells with this genetic change would then expand clonally to develop papillomas by the action of promoters. A further genetic change is required for the irreversible progression of papillomas to carcinomas, a progression phase distinguishing between benign tumor and malignant tumor. By using the induction of hepatocellular carcinoma (rat liver cancer) as a model, Farber (46, 47) and Farber and Sarma (49) have demonstrated that each of the three phases (initiation, promotion and progression) may involve many steps; see also references 28, 48, 121, 127.

Initiation

This phase is characterized by an irreversible alternation in the genetic expression of a stem cell that produces a heritable change in the cell's phenotype. This genetic change may either be a gene mutation or other genetic changes such as some chromosomal aberrations. For example, Balmain et al (7) have reported that the H-ras gene has been mutated to an activated form in both papillomas and carcinomas. This would provide support for the hypothesis that point mutation of the ras oncogene is a key event in the initiation process of mouse skin carcinogenesis. A direct and strong piece of evidence implying point mutation of the ras gene in initiation was provided by Barbacid and coworkers (see 8, 163, 164). By injecting NMU (N-nitrosomethylurea) into the breast of female rats, Zarbl et al (163, 164) found that the NMU molecule binds with the ras DNA, thus inducing a base change from adenosine to guanosine $(G \rightarrow A)$ at the second nucleotide of the 12th codon of the ras gene. In this case, then, initiation involves a single base change $(G \rightarrow A)$ point mutation of the ras oncogene. Similar results on human cancers have been obtained by Almoguera et al (3) on the human carcinomas of the exocrine pancreas, by Corominas et at (34) on the human benign tumors of the skin and by Burmer and Loeb (23) on the human colon polyps. For example, Burmer and Loeb (23) reported that the mutation at the second nucleotide of codon 12 of the k-ras gene leading to a $G \rightarrow A$ base change might initiate the human colon cancer; Corominas et al (34) reported that the mutation at the second nucleotide of codon 61 of the H-ras gene leading to a $A \cdot T \rightarrow T \cdot A$ change might initiate the human benign tumors of the skin (keratoacanthomas).

In general, the initiation process would involve irreversible genetic changes, is additive in its effects and does not exhibit a readily measurable threshold or maximal response within the limits of toxicity. A chemical or agent is called an initiator if it can induce initiation for the process of carcinogenesis.

Promotion

This phase is reversible and is the phase which facilitates clonal expansion of initiated stem cells. Some promoters such as phorbol ester appear to act through a membrane receptor (protein kinase C) while others such as benzoyl peroxide may act through a free radical mechanism (29, 83, 116). The mechanisms of promotion may either be direct or indirect involving many epigenetic factors. The effects of some skin tumor promoters on initiated cells may appear to be mimicking some growth factor, thus exerting a direct effect. On the other hand, modifications in cell-to-cell communication through gap junctions (114, 142–145) and stimulation of differentiation of non-initiated cells are important indirect epigenetic mechanisms. Some genetic changes such as gene amplification and chromosomal changes may also play a role in tumor promotion (46–49, 125–127).

The following summarizes some basic results about promotion:

i. Effects of promoters are in general reversible and not additive.

ii. Promotion alone is incapable of affecting carcinogenesis; promotion is effective only following initiation.

iii. Most promoting agents are not mutagenic.

iv. The effects of promotion exhibit a measurable threshold and a maximal response usually short of toxicity.

v. Antioxidants and other inhibitors of active oxygen formation inhibit tumor promotion *in vivo*.

REMARK   Slaga (125–127) noted that promotion might consist of at least two stages; he has thus proposed stage-I promotion and stage-II promotion. These problems remain to be settled, however. (See 65).

Progression

This phase involves development of metastatic tumor cells, formation of clumps of tumor cells of various sizes, migration of clumps through the circulatory system, arrest of clumps at distant sites and development of metastatic foci at distant sites (91, 109, 139). Recent biological studies have indicated that gene mutations or genetic changes are also required for the generation of metastatic tumor cells; yet at a distant site whether or not metastatic foci are formed depends on if the genes are turned on or off by hypomethylation of DNA (73, 90).

1.3.3   The Impact of Environment

We have demonstrated that under normal environmental conditions, carcinogenesis is a multistep sequential process involving at least two steps. This is often the case if the cancer gene carrying cells are surrounded by a large number of normal cells and if the cancer genes are not linked to strong promoters or enhancers (9–11, 46–49). Recent biological experiments have shown that if the normal surrounding cells are killed or removed or if the cancer genes are linked to a strong promoter or enhancer, then the process of carcinogenesis can be accomplished by a single cancer gene, resulting in a one-stage model of carcinogenesis. For example, Spandidos and Wilkie (128) have shown that tumorigenic conversion of normal rodent stem cells can be achieved by a single activated human oncogene (ras) if it is linked to a strong promoter. Land et al (86) reported that if the surrounding normal cells were killed by neomycin, then a single oncogene (ras) was sufficient to induce tumorigenic conversion of normal rat embryo fibroblast cells. Similar results have been reported by Detto, Weinberg and Ariza (39) on mouse keratinocytes for H-ras oncogene and by Bignami et al (15) on mammalian cells for the viral oncogenes myc, ras and src.

The aforementioned experimental results imply that the process of carcinogenesis and the number of stages depend not only on the cancer genes but also on the environmental conditions. Since different individuals in the population may be subjected to different environmental conditions, it is anticipated that different individuals may involve different pathways and/or stages for carcinogenesis. That is, one may expect a mixture of different models of carcinogenesis in the population. (For some more detail, see Chapter 5).

Also, since different environments may affect different cancer genes which share some common properties (see Section 1.5), the same cancer tumor may arise from different pathways. Because of the changing environments, for the same individual the same type of cancer tumors may arise from different pathways, leading to models of carcinogenesis involving multiple pathways (For more detail, see Chapter 4).

1.4 GENETIC CHANGES IN CARCINOGENESIS

As we have demonstrated in the previous section, genetic changes are the dynamic part of the process of carcinogenesis. These genetic changes include deletion, gene mutation, reciprocal chromosomal translocation, gene amplification and other genic or chromosomal aberrations. Given below are some specific examples relating these genetic changes to certain cancers.

Deletion

In human beings, deletion of chromosome 13q contributes to the development of retinoblastoma (26, 27, 41, 55, 56, 61, 67, 80, 152), osteosarcoma (55, 59, 152), small cell lung carcinoma (63, 155); breast cancer (59, 89, 115, 138) and stomach cancer (115); deletion of chromosome 11p contributes to the development of Wilm's tumor (51, 66, 81, 110, 153), Rhabdomyosarcoma (59), hepatoblastoma (59), hepatocellular carcinoma (115), familial adrenocortical carcinoma (66), bladder cancer (115) and breast cancer (59, 115); deletion of chromosome 3p contributes to the development of renal carcinoma (59, 82) and small cell lung cancer (115, 155); deletion of chromosome 1p contributes to the development of neuroblastoma (53, 115), endocrine neoplasia (98), phaeochromocytoma (115) and medullary thyroid carcinoma (115); deletion of chromosome 17p contributes to the development of small cell lung cancer and colon cancer (147, 155). Recently, deletion of chromosome 18 has also been reported in human colon cancer (147).

Gene Mutation

Gene mutations resulting from the base change at the 12th codon, the 13th codon, the 59th codon, and the 61st codon of the ras gene have been implicated in many human and animal tumors (3, 7, 8, 18–19, 23, 34, 50, 92, 163–164). As an example, the point mutation resulting from a $G \rightarrow A$ base change at the second nucleotide of the 12th codon of the ras oncogene is responsible for the initiation of breast cancer in rats induced by NMU (163, 164). For human beings, Burmer and Loeb (23) reported that a point mutation involving a single base change $G \rightarrow A$ at the second nucleotide of codon 12 of the k-ras gene contributed to the initiation of benign colon polyps; Corominas et al (34) reported that a point mutation involving a $A \cdot T \rightarrow T \cdot A$ base change at the second nucleotide of codon 61 of the H-ras gene contributed to the initiation of human benign tumors of the skin.

Translocation

It is well documented that reciprocal translocations contribute to human leukemia and lymphomas (1, 35, 36, 45, 113, 117). A well-known example among onocologists is Burkitt's lymphoma which involves translocations between chromosome 8 and chromosome 14, or chromosome 2 or chromosome 22 (t(8:14) (q24;q32), t(8;2) (q24;p12) and t(8;22) (q24;q11)). (t(8:14) was found in 80–90% of Burkitt lymphoma.) The breakpoint at chromosome 8 is at q24 while the breakpoints at chromosomes 14, 2 and 22 are at q32, p12 and q11 respectively. Since the myc oncogene is located on the translocated portion of chromosome 8, in the translocations t(8:14) (q24;q32), t(8:2) (q24;p12), t(8:22) (q24;q11), the myc oncogene on chromosome 8 is then moved to the vicinity of one of the three immuno-globulin loci on chromosome 14 (Ig H), chromosome 2 (Ig K) and chromosome 22 (Ig λ), thus activating the myc gene; for more detail, see 35, 36, 45. Another example which has been constantly observed in 90% of the chronic myelogeneous leukemia (CML) is a reciprocal translocation t(9:22) (q34:q11) between chromosomes 9 and 22.

(This translocation was known in the past as the Philadelphia chromosome.) This translocation fuses the proto-oncogene c-abl in chromosome 9 with a newly recognized genetic locus known as bcr (breakpoint cluster region) of chromosome 22. This fusion creates a chimeric protein that includes the functional domain of the c-abl gene product, but whose enzymatic activity is more robust than that of a normal gene product.

Gene Amplification

Gene amplification is one of the genetic alterations which activates cellular oncogenes (2, 30, 44, 58, 68). At least six oncogenes have been found to be amplified and the degree of gene amplification varies up to many hundredfold over the single haploid copies found in normal stem cells. c-myc has been found to be amplified about 30–50 fold in some human colon carcinoma cells, about 16–32 fold in some human promyelocytic leukemia cells, about 20–70 fold in some human small-cell lung cancer variants and about 10 fold in some human breast carcinomas; N-myc has been found amplified in about 15 to over 100 fold in human neuroblastomas; c-myb has been found amplified in about 10 fold in some human colon carcinoma; c-abl has been found amplified in about 10 fold in some human chronic myelogeneous leukemia cells. c-Ki-ras and N-ras have also been found amplified in some tumor cells. (2, 58).

Amplified oncogene may either occupy in DM (Double minute) chromosomes or HSR (Homogeneous Staining Regions); examples include human colon carcinoma cells, c-Ki-ras in some mouse adrenal tumor cells and N-myc in human neuroblastoma cells. DM chromosomes are fragments of chromosomes without centromeres so that during cell division they will either be lost or integrated into HSR regions in regular chromosomes. HSR regions may break up to give DM's. Occasionally, amplified oncogenes may also occupy other regions than DM's or HSR; it may also accompany chromosomal translocations or chromosomal rearrangements.

In most cases, gene amplification may not initiate the process of carcinogenesis since in most cases not all tumor cells have amplified genes. However, two recent studies suggested that in some cases it might be possible for gene amplification to initiate the carcinogenesis process. In fact, using NIH3T3 transfectant DHFR/G-8 cell lines, Hung, Yan and Zhao (68) reported that amplification of the proto-neu oncogene facilitated oncogenic activation by a single point mutation. Working on rat liver tumors, Chander, Lombardi and Locker (30) reported that c-myc gene amplification could be induced by a choline-devoid diet without adding carcinogens.

Repair Defects and Chromosome Breakage Syndrome

When stem cells are exposed to radiation or chemicals, the exposure may induce DNA damage and/or chromatid breaks, leading to lesions. Normal stem cells contain a DNA repair system to repair these lesions. Mutations result if DNA damage is mismatchly repaired (error-prone repair, 113, 130, 141). In the inherited human disease xeroderma pigmentosum (xp), for example, Sirover et al (123) reported that for patients who are homozygotes for the xp gene, the DNA repair regulation system was defective; for these patients, therefore, exposure to ultraviolet light would lead to initiation of carcinogenesis for the human skin. Examples for chromosome breakage syndrome in human beings include ataxia telangiectasia (at), Bloom's syndrome and Fanconi's anemia. For example, Langlois et al (87) reported high frequencies of somatic crossing over and sister chromatid exchange in erythroid precursor cells in Bloom's syndrome patients, predisposing to leukemia and other human cancers. Cavenee et al (26, 27) reported somatic chromatid crossing over and nondisjunction for the second event in hereditary retinoblastoma cases, see also reference 61.

1.5 CANCER GENES

To understand the disease of cancer, it is essential to be familiar with cancer genes and how these genes function.

There are basically three types of cancer genes which are responsible for carcinogenesis: Oncogenes, antioncogenes (or suppressor genes) and accessory genes (or modifier genes). Oncogenes and antioncogenes are genes directly responsible for carcinogenesis while accessory genes are indirectly related to carcinogenesis whose functions would affect the rates of initiation, and/or cell proliferation and differentiation of initiated cells (and/or normal stem cells). Given in Table 1.1 is a brief comparison between oncogenes and antioncogenes. For human beings, many of the cancer genes have been mapped on the 23 pairs of chromosomes (22, 124).

Table 1.1 Comparison Between Oncogenes and Antioncogenes

Features	Oncogenes	Antioncogene (Suppressor gene)
Function	Activated	Inactivated
Host Genetic	Nonhereditary	Hereditary or Nonhereditary
Cellular Genetic	Dominant	Recessive
Cytology	Amplification, Translocation or Point mutation	Loss of heterozygosity, deletion, or point mutation

1.5.1 Oncogenes

Oncogenes are highly preserved regulatory genes which regulate cell proliferation and differentiation. The normal form of oncogenes has been referred to as protooncogenes. When protooncogenes are activated or mutated to become oncogenes, normal control of cell growth is unleashed, leading to the cascade of carcinogenesis (8, 14, 18–20, 22, 38, 59, 62, 76–77, 92, 96–97, 101, 113, 131, 141, 148, 150–151).

Oncogenes were first identified in the early 1970 in Rous sarcoma virus (RSV), which causes cancer in chickens. (This oncogene has been referred to as v–src.) Since then many other virus oncogenes and cellular copies (called c–oncogenes) have been discovered. It appears that many v–oncogenes (viral oncogenes) are truncated versions of their cellular counterpart (20, 96–97). In human beings, the most common oncogenes are the ras oncogene family (H–ras, Ki–ras and N–ras) whose encoded proteins function and reside in the cytoplasm and membrane and the myc oncogene family (c–myc, N–myc, L–myc) and the oncogene myb whose encoded protein function in the nucleus. These oncogenes have been isolated and cloned; also the DNA base sequences for these oncogenes have been determined (8, 18–20, 22, 38, 92). There are at least four avenues by means of which proto-oncogenes can be activated to become oncogenes. These avenues include: point mutation, chromosomal translocation, gene amplification and insertional carcinogenesis by retrovirus. The first three avenues have been discussed in Section (1.4) while activation through retrovirus will be discussed in Section (1.7).

Up to 1989, more than 57 viral and cellular oncogenes have been reported. These oncogenes together with their origin, function and chromosomal sites have been tabulated by Burck, Liu, and Larrick (22) in the appendix of their book. According to the functions and locations of its encoded proteins, these oncogenes have been classified into the following six categories:

1. Oncogenes whose encoded proteins are related to Tyrosine–specific protein kinases (Cytoplasmic). Oncogenes in this class include src, ros, fes, yes, fgr, abl, raf, mil. Most of the proteins encoded by these oncogenes reside in the plasma membrane.
2. Growth Receptor Related Oncogenes. This class includes oncogenes erb-B and fms whose encoded proteins reside in plasma and cytoplasmic membranes and are related to potential protein kinases. The oncogene erb–B is known to affect the EGF receptor while the oncogene fms is known to affect the C5F–1 receptor.
3. Growth Factor Related Oncogene. A well–known example is the secreted oncogene sis which controls PDGF.
4. The ras–family oncogenes include H–ras, Ki–ras and N–ras. These oncogenes encode GTP-binding proteins which function and reside in the plasma membrane.
5. Nucleus protein related oncogenes include c–myc, N–myc, L–myc, myb, P53, fos and ski. The encoded proteins of these oncogenes reside in the nucleus and are either gene regulators or enhancers.
6. Oncogenes whose functions are yet unknown and cannot be classified into any of the above five classes. This class includes oncogenes ets, rel, fim-1, fim-2, int-1 and mel, among many others.

Relative to the function sites of its encoded proteins, oncogenes can also be classified either as cytoplasmic oncogenes whose encoded proteins function and reside in the cytoplasm or plasma membrane, or as nuclear oncogenes whose encoded proteins function and reside in the necleus. Examples of cytoplasmic oncogenes include viral oncogene polyoma middle T and cellular oncogenes ras, src, erbB, neu, ros, fms, fes/fps, yes, mil/raf, mos, abl; examples of nuclear oncogenes include viral oncogenes SV40 large T, Polyoma large T, Adenovirus ElA and cellular oncogenes myc, myb, N-myc, P53, ski and fos. One of the major differences between cytoplasmic oncogenes and nuclear oncogenes appear to be that the proteins encoded by cytoplasmic oncogenes are altered structurally as compared with proteins encoded by protooncogenes while most of the proteins encoded by nuclear oncogenes do not seem to be much different from their normal proteins constitutively (22, 148). This difference leads to the result that the functions of cytoplasmic oncogenes are quite different from those of nuclear oncogenes; recent biological results establish that these two types of oncogenes complement each other and collaborate in inducing tumorigenic conversion of normal stem cells (4, 33, 57, 85, 96–97, 118, 122, 148, 151).

The Immortalization Class and the Transformation Class of Oncogenes

The functions of oncogenes within each group, (cytoplasmic or nuclear), although not precisely the same, share some common properties. Nuclear oncogenes are in general strong in inducing immortalization but appear to be weak in their ability to induce anchorage independence of fibroblasts. In contrast to nuclear oncogenes, cytoplasmic oncogenes are generally weak in their ability to immortalize cells but strong in their ability to promote anchorage independence of fibroblasts (22, 148). Because of these distinctive functions for cytoplasmic oncogenes and nuclear oncogenes, the class of cytoplasmic oncogenes has been referred to as transforming group while the class of nuclear oncogenes as immortalizing group. There are exceptions to these functional differences, however. For example, the SV40 large T oncogene, which, through the action of a single protein, can induce both immortalization and anchorage independence; SV40 large T oncogene has thus been referred to as a "dual" oncogene (25).

Cooperation Between Oncogenes

Land, Parada and Weinberg (85) reported that cotransfection of the oncogenes ras and myc would induce tumorigenic conversion of rat embryo fibroblasts in culture; yet the ras or myc gene alone would not convert the rat embryo fibroblasts into tumor cells under usual conditions. These and many other laboratory works (4, 33, 57, 85, 96–97, 118, 122, 148, 151) underline a general principle which implies that nuclear oncogenes would in general collaborate effectively with cytoplasmic oncogenes in malignant transformation of previously normal cells while under usual conditions, a cytoplasmic oncogene or a nuclear oncogene alone is unable to induce full transformation. There are exceptions to this rule, however; for example, SV40 large T oncogene alone can induce tumorigenic transformation of normal cells; Spandidos and Wilkie (128) have also reported that a ras oncogene alone, when connected with a promotor or enhancer, is able to induce tumorigenic conversion of rat embryo fibroblasts (see also 15, 25, 39, 86, 151).

1.5.2 Antioncogenes (Suppressor Genes)

Antioncogenes (or suppressor genes) are genes which suppress the expression of oncogenes or other genes so that its inactivation or deletion would lead to the carcinogenesis cascade (75, 80, 149, 151). Thus, unlike oncogenes which act in a dominant fashion, antioncogenes are recessive genes so that only homozygotes or hemizygotes for the gene give rise to the cancer phenotype. Thus, for cancers which involve only a single antioncogene, it takes two consecutive events leading to the cancer phenotype. In this model, the first event is the deletion or mutation of one of the alleles, which happens either in germline cells or in somatic cells; the second event is the mutation or loss by nondisjunction or chromatid exchange or somatic crossing over or deletion of the second allele in the somatic cell leading to homozygosity or hemizygosity for the antioncogene (26–27, 41, 51, 61, 78, 81–82, 153).

In human beings, antioncogenes have been located in chromosomes 1p, 3p, 5q, 11p, 13q, 17p, 18q and 22. In human beings, well-known examples of antioncogenes include the childhood cancers retinoblastoma, the Wilm's tumor and colon cancer. For retinoblastoma, there is a Rb gene in chromosome 13q14 (26–27, 41, 55–56, 61); for Wilm's tumor, there is a Wm gene in chromosome 11p13 (51, 61, 81, 110, 153); and for colon cancer there is a FAP gene in chromosome 5q and also possibly antioncogenes in chromosomes 17p and 18q (147).

The location of antioncogenes (or suppressor genes) and the demonstration of their suppressing effects have been achieved by hybridization studies between normal cells and cancerous cells. In human beings, hybrids between normal cells and cancer cells often do not produce tumors in nude mice although the cancerous parental cells do; however, these hybrids regain their tumorigenicity if chromosome 11 is lost, indicating that chromosome 11 contains suppressor genes (110). Barrett et al (12) reported that hybrids between normal Syrian hamster embryo (SHE) cells and tumorigeneous SHE cells which were induced by v–H–ras plus v–myc oncogenes had lost their tumorigenicity; yet these hybrids still expressed the ras and myc RNA and high levels of the mutated form of the $p21^{ras}$ protein, indicating that the loss of tumorigenicity was not due to the loss or lack of expression of the oncogenes. When these hybrid cells were passaged, anchorage–independent and tumorigenic variant cells arose at a low frequency in the populations. Karyotype analysis indicated that hybrids which were suppressed for tumorigenicity and anchorage–independence contained normal karyotypes as normal cells;

yet hybrids which re–expressed tumorigenicity had a nonrandom loss of chromosome 15. These and many other results (119) suggest that normal antioncogenes suppress the effects of oncogenes; this in turn suggests that for the cancer phenotype many other events may accompany the inactivation of antioncogenes. For example, Chaum et al (31) reported that for the retinoblastoma phenotype, the deletion of the Rb locus (del(13)(q12 → 14)) was usually accompanied by consistent chromosomal abnormalities which included trisomy for chromosome 1q, the presence of i(6p) which contributes two extra copies of 6p and loss of 6q. (For more results, see 119.) Recent studies showed that loss or mutational inactivation of the normal retinoblastoma gene at chromosome 13q14 were associated with a variety of human cancers which include retinoblastoma, osteosarcoma, breast cancer, small cell lung cancer (SCLC) and stomach cancer (26–27, 41, 55–56, 59, 61, 63, 67, 80, 115, 152, 155). Similarly, deletion of the Wilm tumor gene at chromosome 11p13 has been shown to be associated with Wilm's tumor, Rhabdomyosarcoma, hepatoblastoma, hepatocellular carcinoma, familial adrenocortical carcinoma, bladder cancer and stomach cancer (51, 59, 66, 81, 110, 115, 153).

1.5.3 Accessory Cancer Genes

Accessory cancer genes are genes which segregate in Mendelian fashion and are highly predisposable to certain cancers (79). Thus, unlike oncogenes and antioncogenes, accessory cancer genes are genes which relate to cancers indirectly by increasing mutation rates of oncogenes and antioncogenes, and/or by facilitating cell proliferation of intermediate cells and/or cancer progression. Examples of accessory cancer genes in human beings include xeroderma pigmentosum (xp), ataxia telangiectasia (at), Fanconi's anemia (fa) and Bloom's syndrome. Consider for example the human inherited disease xeroderma pigmentosum for skin cancer. It has been shown (124) that for patients homozygous for the cancer prone gene xp, DNA repair for damages by ultraviolet light is defective, leading to increased mutation rate for the initiation process in skin cancer. As another example, consider Bloom's syndrome which involves an autosomal recessive gene for genetic disorder of growth. For individuals who are homozygous for this gene, Langlois et al (87) reported increased somatic crossing over, increased sister-chromatid exchanges and chromosome breaks and rearrangement in erythroid precursor cells, leading to increased frequency in a variety of cancers including lymploid, myeloid leukemia, lymphomas, squamous cell carcinomas and adenocarcinomas.

1.6 THE FUNCTIONS OF ONCOGENES AND ANTIONCOGENES

Genes are DNA molecules which produce RNA which in turn encodes proteins or enzymes. Along this pathway, oncogenes and antioncogenes exert its effect through its protein products. An immediate question is then: How do oncogenes and antioncogenes act at the molecular level? Although many aspects remain elusive, the past 10 years have revealed many of the mysteries. In this section I shall proceed to give a very brief account of how oncogenes and antioncogenes act at the molecular level.

1.6.1 The Function of Oncogenes at the Molecular Level

The Autocrine Theory of Carcinogenesis

Oncogenes function to release the cells from regulated growth control. As such, a major avenue is to release the cells from their dependence on growth factors for cell proliferation. The mechanisms leading to the independence of cell proliferation on growth factors contribute to the autocrine theory of carcinogenesis. For cytoplasmic oncogenes at least three avenues have been revealed leading to the autocrine status of cells (8, 20, 22, 60, 76, 84, 148, 160).

i. The first avenue is the direct growth factor protein encoded by oncogenes. Examples of this avenue include the sis oncogene which encodes a protein almost identical to the platelet-derived growth factor (PDGF), and the oncogenes hst and int-2 which encode proteins belonging to the fibroblast growth factor (FGF) family (22, 101).

ii. The second avenue is related to the growth factor receptor. Examples of this avenue include the oncogenes erbB, neu and fms. To illustrate how the oncogene functions by way of the growth factor receptor, a well documented example is the oncogene erbB encoding a protein which is a truncated version of the epidermal growth factor (EGF) receptor. The normal EGF receptor spans over the cell membrane with the EGF binding region extending outside the cell while the catalytic domain (protein kinase domain) extends inside the cytoplasm of the cell. At the external application of epidermal growth factor (EGF), the EGF binds with the extracellular EGF binding domain; such an extracellular binding changes the receptor's conformation so that a substrate protein (a p42 protein) in the cytoplasm then binds with the protein kinase domain of the EGF receptor, leading to the phosphorylation of this cellular substrate protein (p42). The phosphorylation of p42 appears to be the first step in the pathway delivering to the normal cell the signal to divide. The erbB oncogene protein, being a truncated version of the EGF receptor, lacks the EGF receptor's EGF–binding region so that signals to phosphorylate the cellular substrate protein are being constantly delivered even without application of extracellular epidermal growth factors.

iii. The third avenue to the autocrine status is by way of signal transduction. An important example for this avenue is the H–ras oncogene. The proto–oncogene H–ras encodes a p21 protein which is a G–type protein. p21 binds with guanosine triphosphate (GTP) to yield excited p21:GTP. The excited p21:GTP activates the enzyme adenyl cylase which catalyzes production of phosphodiesterase (PDE); PDE further catalyzes the hydrolysis of phosphatidylinositol 4,5–biphosphate (PIP_2) to yield second messengers diacylglycerol (DG) and inositol 1, 4, 5–triphosphate (IP_3), leading to synthesis of DNA and cell division. In activating adenyl cyclase, p21:GTP is hydrolyzed to produce guanosine diphosphate (GDP), p21 and P (phosphate). To understand the basic function of the ras oncogene, observe that a H–ras oncogene again encodes a G–type protein $p21^{ras}$ but this $p21^{ras}$ is different from the original p21 protein by one amino acid. (In the G to A point mutation at the second nucleotide of the 12th codon, amino acid glycine is replaced by valine.) The new $p21^{ras}$ is unable to hydrolyze GTP to yield GDP, thus p21:GTP remains to be excited, leading to continuous production of adenyl cyclase.

The above two examples, the erbB oncogene and the ras oncogene, reveal a general rule for cytoplasmic oncogenes. That is, cytoplasmic oncogenes exert their functions through structurally altered protein products.

In contrast to cytoplasmic oncogenes, the nuclear oncogenes such as myc and fos seem to function by over expression of the gene yielding an increased level of its proteins; thus the encoded proteins of nuclear oncogenes are constitutively the same as its proto-oncogene products. These oncogene proteins generally bind with DNA leading to replication of DNA; alternatively, these proteins may influence biosynthesis of mRNA, by means of which they mobilize the expression of a bank of cellular genes whose products are critical to growth and differentiation (20, 22, 92, 148, 154). As to how these mechanisms work the major steps remain elusive.

Second–Messenger Pathways and Phosphorylation of Mitogenic Agents

At the molecular level, there are basically two signal pathways by means of which cytoplasmic oncogenes promote DNA synthesis and cell proliferation: Phosphorylation of substrate proteins which serve as intracellular mitogens, and production of second messengers and activation of the inositol lipid cycle (13, 22, 108). The oncogenes ras, sis, abl, mos and src appear to function through second messengers while the oncogenes erbB, yes, fgr, fps, fes, mil and raf seem to function by way of phosphorylation of substrate proteins which serve as intracellular mitogens.

    i.   The phosphorylation pathway is to phosphorylate certain proteins which serve as intracellular mitogenic agents, either based on tyrosine phosphorylation by the action of tyrosine kinase, or based on serine phosphorylation by the action of serine kinase. (See Section 1.5.1 for the corresponding oncogenes.) In these processes, the phosphate source comes from ATP (adenosine triphosphate); the action of tyrosine kinase adds a phosphate to the tyrosine residue of the target protein while reducing ATP to ADP (adenosine biphosphate). Since phosphate is the energy source, phosphorylation of the protein then carries out the effects of external growth factors such as EGF and insulin to the nucleus to signal DNA synthesis. A specific example along this pathway is the erbB oncogene which is related to the PDGF; for details, see Section (1.6.1)(a).

    ii.   The second–messenger signal pathway occurs either by catalyzing the hydrolysis of PIP_2 (phosphatidyl inositol 4,5–biphosphate) to yield two second messengers DG(Diacylglycerol) and $IP_3(1, 4, 5)$ (Inositol 1, 4, 5–triphosphate), or by activating the inositol lipid cycle to yield DG and $IP_3(1, 4, 5)$. Now, $IP_3(1, 4, 5)$ functions to mobilize calcium from intracellular stores while DG activates protein kinase C, one function of which is to increase intracellular pH by switching on a N_a^+/H^+ exchanger. Since the onset of DNA synthesis requires two ionic events, the increase in pH and calcium, the production of DG and $IP_3(1, 4, 5)$ then leads to promotion of DNA synthesis. The second messengers DG and $IP_3(1, 4, 5)$ may also affect DNA synthesis by either entering the inositol lipid cycle or by yielding potent mitogens or potent growth factors; for example, two potent mitogens prostaglandin $E_1(PGE_1)$ and prostaglandin $F_{2\alpha}(PGE_{2\alpha})$ are products of arachidonic acid which in turn is a product of DG; the potent second messenger $I_{ns}1, 3, 4 - P_3$ (Inositol 1, 3, 4–triphosphate) is a product of $I_{ns}1, 3, 4, 5 - P_4$ (Inositol 1,3,4,5–tetrakisphosphate) which in turn is a product of $IP_3(1, 4, 5)$. Other functions of protein kinase C are to catalyze the production of PIP_2 from PIP (PI 4–phosphate) and the production of PIP from PI (phosphatidylinositol), thus activating the inositol lipid cycle.

Examples of oncogenes which function by second–messenger pathway include the oncogenes ras, abl, ros and src. The oncogenes abl, ros and src may function by catalyzing the processes PI $\rightarrow$ PIP $\rightarrow$ PIP_2 and hence promote DNA synthesis since PIP_2 is a precursor of the second messengers DG and $IP_3(1,4,5)$. The ras gene functions by activating adenylcyclase which catalyzes the production of PDE (phosphodiesterase) while PDE catalyzes the hydrolysis of PIP_2 to yield DG and $IP_3(1,4,5)$; see Section (1.6.1)(a).

1.6.2 The Action of Antioncogene at the Molecular Level

For understanding the action of antioncogenes at the molecular level, many recent studies have been directed to the retinoblastoma (Rb) gene at 13q14 which is a 190-kilobase stretch (37, 42, 56, 67, 88, 149, 159). (To date the retinoblastoma gene appears to be the only gene which is manageable in the laboratories.) In 1986, the normal Rb locus was isolated by Lee and coworkers (56, 88) and the structure of its encoded protein (a Mr 105,000 phosphoprotein to be referred to as the 105p-Rb protein) sequenced (88). This protein appears to regulate growth of normal cells in a number of organs including the retina tissue. Subsequently it was discovered that the 105p-Rb protein formed a complex with the protein of the oncogene ElA (149, 151, 159); this process interferes with the normal function of the Rb gene product (e.g., the 105p-Rb protein) and therefore, like loss of the Rb gene, leads to uncontrolled growth and tumor formation (149, 151, 159). After the above discovery, it has also been discovered that the 105p-Rb protein forms a complex with the encoded proteins of two other oncogenes; the SV40 antigen oncogene and the human papillomavirus 16E7 oncogene (37, 42). These three oncogenes (ElA, SV40 and 16E7) appear to be nuclear oncogenes and share some common properties with the myc oncogene. (The function of myc can be replaced by that of ElA.) One may perhaps generalize the above mechanism to the myc oncogene. Recently, Erisman, Scott and Astrin (43) found in a subset of human colon cancers that the elevated level of c-myc expression was frequently correlated with the loss of the antioncogene Fap at 5q21-q22 and none of the cells with normal c-myc level showed loss of the Fap gene. This result may suggest some interaction between the myc oncogene and the antioncogene FAP; it is not clear at this moment how the interaction works, however.

1.7 INSERTIONAL CARCINOGENESIS AND ACTIVATION OF ONCOGENES BY A RETROVIRUS

A retrovirus can activate oncogenes and hence induce cancers in the host (human beings or animals). The basic mechanism by means of which a retrovirus activates oncogenes in the host has been called "Insertional Mutagenesis" or "Insertional Carcinogenesis."

A retrovirus is a virus made up of RNA and a protein coat. When a retrovirus invades a host, its RNA enters a cell of the host, leaving the protein coat outside the cell. By the work of the enzyme reverse transcriptase, the viral RNA in the cell then produces viral DNA; next the viral DNA enters the nucleus and is inserted into a chromosome of the host genome. The viral RNA (and hence the viral DNA) contains a unit called LTR (Long Terminal Repeat) which activates the viral oncogene if the virus contains v–oncogenes. (These viruses are called acute transforming retroviruses.) In these cases, the activated viral oncogenes then lead to the carcinogenesis cascade of the host individual. If the retrovirus does not contain oncogenes (these retroviruses are called chronic retroviruses; examples include HTLV1 and human papilloma virus), the LTR unit of the virus then serves as a promoter or enhancer of a cellular oncogene, resulting in an increased transcription of the cellular oncogene. The insertion of viral DNA may also elicit point mutations within an adjacent cellular gene (Westway et al 156). Oncogenes which have been activated by insertional mutagenesis of a retrovirus include c–erbB, c–myb, c–myc, c–H–ras and c–mos, among many others.

1.8 SOME MATHEMATICAL MODELS OF CARCINOGENESIS

By scrutinizing and integrating the biological findings of cancer biology given in previous sections, some useful models of carcinogenesis can be formulated. In this section we thus present very briefly some of these models. The mathematical theories of these models will be developed and illustrated in detail in Chapters 3–6.

The Multievent Model of Carcinogenesis and Extensions

This model was first proposed by Chu (32) in 1985 as an extension of the classical Armitage–Doll multistage model of carcinogenesis. A stochastic version of this model has been derived by Tan, Chu and Brown (134). This model assumes that cancer tumors develop from stem cells by going through a fixed number k $(k > 1)$ of heritable genetic changes. It differs from the classical Armitage–Doll model in that the intermediate cells are assumed to be subjected to stochastic birth-death processes for cell proliferation and cell differentiation. A schematic presentation of this model is given in Figure 1.2.

The MVK Two-Stage Model of Carcinogenesis and Extensions

If $k = 2$, the above multi-event model reduces to the two-stage model considered by Moolgavkar and Venzon (105) and Moolgavkar and Knudson (106), see also references 103–104, 133, 135. It appears that for cancers which involve only a single anti-oncogene, this model provides a mathematical description of the biological mechanism of carcinogenesis. It is also appropriate in tissues when tumorigenic conversion of normal stem cells involve only immortalization and transformation.

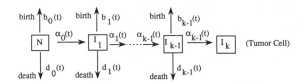

$N = I_0$ = Normal Stem Cell, I_i = ith Intermediate Cell,

$i = 1, \ldots, k - 1$, $I_k = T$ = Tumor Cell,

$\alpha_i(t)$ = Mutation rate of I_i cell at time t,

$b_i(t)$ = Birth rate of I_i cell,

$d_i(t)$ = Death rate of I_i cell, $i = 0, 1, \ldots, k$.

Figure 1.2 The multievent model of carcinogenesis.

The Mixed Models of Carcinogenesis

As we have demonstrated in Section (1.3.3), the process of carcinogenesis and the number of stages depend not only on cancer genes but also on the environmental conditions. Because different individuals in the population may be subjected to different environments, it is anticipated that different individuals may undergo a different pathway and/or a different number of stages for cancer tumor development. These considerations have led Tan (132) and Tan and Singh (136) to develop mixed models of carcinogenesis.

Mixed models of carcinogenesis also arise in cancers which involve both hereditary and non-hereditary cases. This is true in cancers which involve a single antioncogene since in these cases the first event (mutation or deletion of one of the antioncogene) can occur either in germline cells (hereditary) or in somatic cells (nonhereditary) while the second event always occurs in somatic cells (see 26–27, 61, 80, 105–106, 136).

The Multiple Pathway Models of Carcinogenesis

As we have illustrated before, the process of carcinogenesis depends on both the cancer genes and the environmental conditions. Because of the changing environments and because of the fact that oncogenes in the same class (immortalized class or transformation class) share some common properties, it is anticipated that the same tumor may arise from different pathways. This leads to multiple pathway models of carcinogenesis. As shown by Medina (102) and Tan and Chen (137), multiple pathway models of carcinogenesis are quite common and provide a logical explanation to many inconsistent epidemiological data of cancers (see Chapter 4 for details).

1.9 THE ARMITAGE–DOLL MULTISTAGE MODEL AND COMMENTS

Even as early as 1950, it was recognized that for some cancers such as lung cancer, the cancer incidence rate increased very sharply as age increases, giving rise to a log–log linear curve. That is, log incidence is linearly related to log age. This log linear curve prompted Armitage and Doll (5, 6) to propose a multistage model for carcinogenesis. This model assumes that cancer develops from a single stem cell by going through a series of irreversible, heritable, mutation–like events. It differs from

Chu's multievent model in that the Armitage–Doll model completely ignores cell proliferation and cell differentiation of intermediate cells.

If the mutation rates are very small and are independent of time, the incidence function of cancer for the Armitage–Doll multistage model is approximately given by:

$$\lambda(t) \propto (t - w)^{k-1} \tag{1.1}$$

where k is the number of stages and ω is a fixed positive number for the growth of the tumor (for a proof, see Chapter 6).

By (1.1), $\log \lambda(t) = c + (k - 1) \log (t - \omega)$ where c is a constant. (1.1) provides the basic functional relationship for the Armitage–Doll's multistep model. It has been used quite successively by Doll (40) to describe the relationship between lung cancer and smoking. Figure 1.3 gives the incidence functions of lung cancer versus age and smoking duration. It is apparent that the fittings are quite close. It does not follow, however, that (1.1) implies necessarily the correctness of the Armitage–Doll model since many other models fit at least equally well the above incidence function (74, 106).

Much biological data have indicated that cell proliferations of normal stem cells and immortalized cells are important aspects of carcinogenesis (9–11, 21, 46–49, 72, 94, 99, 100, 107, 127). These observations make the classical Armitage–Doll multistage model biologically not plausible since it completely ignores cell proliferation of normal stem cells and intermediate cells. The following summary provides major objections to the Armitage–Doll multistage model:

i. Incidence curves of many human cancers other than lung cancer cannot be explained by this model. Examples include childhood cancer such as retinoblastoma and Wilm's tumor, breast cancer and Hodgkin's disease (104, 106).

ii. Biological evidence suggests that cell proliferation and cell differentiation are important aspects of carcinogenesis (9–11, 21, 46–49, 72, 94, 99, 100, 107, 127). For example, Borek and Sachs (17), Kalunaga (71), Farber (47) and Farber and Sarma (49) showed that at least one round of cell proliferation was required before the mutational change became permanent. The initiation–promotion–initiation experiments of Hennings at al. (64) and Scherer et al. (120) indicate that before another genetic change occurs, the initiated cells must undergone cell proliferations; see also references 9–11, 45–49.

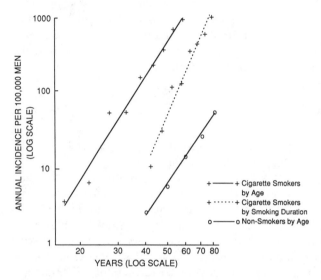

Figure 1.3 Incidence of lung cancer vs smoking. Reproduced with permission from: The Age Distribution of Cancer: Implications for Models of Carcinogenesis, J. Roy. Stat. Soc. A, 1971, vol. 134, pp. 133–166.

iii.   The initiation and promotion experiments of carcinogenesis cannot be explained by the Armitage–Doll multistage model without cell proliferation (104, 121, 127). Similarly, the effects of hormone on endometrian, ovary and breast cancers cannot be explained by the Armitage–Doll model (103–104, 106).

1.10   CONCLUSIONS AND SUMMARY

For providing biological support for the stochastic models of carcinogenesis, in this chapter we present a brief review of the most recent developments of cancer biology. By using results from initiation and promotion studies and by using cell culture studies of Barrett and co-workers on the cancer tumor development of rat tracheal cells and Syrian hamster embryo fibroblasts, it is shown in Section (1.3) that cancer tumors develop from single stem cells by going through a series of genetic changes with intermediate cells subjected to stochastic cell proliferation and differentiation involving an interplay of many genetic, epigenetic and environmental factors. That is, carcinogenesis is a multistep and sequential random process involving genetic changes and stochastic birth-death processes for the fixation of these genetic changes. Further, environmental factors play an important role in determining the pathways and the number of stages in the multistep process of carcinogenesis.

To demonstrate the dynamic roles of genetic changes in carcinogenesis, we give in Section (1.4) some specific examples of genetic changes and illustrate the fundamental roles of these genetic changes in cancer tumor development. These genetic changes include deletion, gene mutation, reciprocal translocation of chromosomes, gene amplification and DNA repair and chromosome breakage syndromes.

For understanding the disease of cancer, we present in Section (1.5) cancer genes and some of its properties and illustrate in Section (1.6) how these cancer genes function at the molecular level. Specifically, there are three types of cancer genes—the oncogenes, the antioncogenes (suppressor genes) and the accessory cancer genes (modifier genes) with the oncogenes and antioncogenes directly responsible for carcinogenesis and with the accessory cancer genes affecting carcinogenesis indirectly. Oncogenes are highly preserved regulatory genes which regulate cell proliferation and differentiation and hence its activation or mutation contributes to the cascade of carcinogenesis. It turns out that some of the encoded proteins of some cytoplasmic oncogenes are growth factors, some are growth factor receptors and some are growth signal transducers or transformers; it follows that the function of these oncogenes is to release the cells from their dependence on growth factors for cell proliferation (autocrine theory). At the molecular level, the cytoplasmic oncogenes function either through phosphorylation of intra-cellular mitogenic agents or through second messenger pathways. Contrary to the oncogenes, the antioncogenes appear to be suppressor genes whose effects are to inhibit cell proliferation and/or to promote cell differentiation so that its loss or inactivation leads to carcinogenesis. Thus, when the protein product of a normal antioncogene is lost or inactivated through forming a complex with the products of oncogenes, the normal control of growth is unleashed leading to uncontrolled growth and tumor formation. This result was illustrated by recent studies on the antioncogene Rb (retinoblastoma) and ElA.

Having presented the most recent development of cancer biology, we then formulate some stochastic models of carcinogenesis as given in Section (1.8) by integrating the biological findings of cancer. These stochastic models will be studied in detail in Chapters 3–6. Finally to conclude this chapter, some comments and criticisms are directed to the classical Armitage–Doll multistage model of carcinogenesis.

REFERENCES

1. Adachi, M. et al., Variant translocation of the bcl-2 gene to immunoglobulin λ light chain gene in chronic lymphocytic leukemia, Proc. Natl. Acad. Sci. USA **86** (1989), 2771–2774.
2. Alitalo, K. and Schwab, M., Oncogene amplification in tumor cells, Advances in Cancer Research **47** (1986), 235–281.
3. Almoguera, C. et al., Most human carcinomas of the exocrine pancreas contain mutant c-k-ras genes, Cell **53** (1988), 549–554.
4. Andeol, A. et al., Both N-ras and c-myc are activated in the SHAC human stomach fibrosarcoma cell lines, Int. J. Cancer **41** (1988), 732–737.
5. Armitage, P. and Doll, R., The age distribution of cancer and multistage theory of carcinogenesis, Brit. J. of Cancer **8** (1954), 1–12.
6. Armitage P. and Doll, R., Stochastic models for carcinogenesis, in "Fourth Berkeley Symposium on Mathematical Statistics and Probability," Univ. California, Berkeley, Cal., 1961.
7. Balmain, A. et al., Activation of the mouse cellular Harvey–ras gene in chemically induced benign skin papillomas, Nature **307** (1984), 658–660.
8. Barbacid, M., Ras genes, Annu. Rev. Biochem **56** (1987), 779–827.
9. Barrett, J. C., Cell culture models of multistep carcinogenesis, IARC Scientific Publication No. 58. "Age–Related Factors in Carcinogenesis." Eds. A. Likhavhev, V. Anisimov & R. Montesano, Lyon, France, 1988.
10. Barrett, J. C., A model for neoplastic development: Role of genetic and epigenetic changes, In "Mechanisms of Environmental Carcinogenesis, Vol II, Multistep Models of Carcinogenesis,", Edited by Barrett, J. C., CRC Press, Boca Raton, 1986..
11. Barrett, J. C. and Fletcher, W. F., Cellular and molecular mechanisms of multistep carcinogenesis in cell culture models, In "Mechanisms of Environmental Carcinogenesis, Vol II, Multistep Models of Carcinogenesis," Edited by Barrett, J. C., CRC Press, Boca Raton, 1986.

12. Barrett, J. C. et al., Oncogene and chemical–induced neoplastic progression: Role of tumor suppression, National Inst. Environmental Health Sciences, NIH, Research Triangle, N. C., 1987 (1987).
13. Berridge, M. J., Growth factors, oncogenes and inositol lipids, Cancer Survey **5** (1986), 413–430.
14. Bishop, J. M., The molecular genetics of cancer, Science **235** (1987), 305–311.
15. Bignami, M. et al., Differential influence of adjacent normal cells on the proliferation of mammalian cells transformed by the viral oncogenes myc, ras, and src, Oncogene **2** (1988), 509–514.
16. Bizzari, J–P., Mackillop, W. J. and Buick, R. N., Cellular specificity of NB/70K, a putative human ovarian tumor antigen, Cancer Research **43** (1983), 864–867.
17. Borek, C. and Sachs, L., The number of cell generations required to fix the transformed state, Proc. Natl. Acad. Sci. USA **59** (1968), 83–85.
18. Bos, J. L., The ras gene family and human carcinogenesis, Mutat. Res. **195** (1988), 255–271.
19. Bos, J. L., Ras oncogenes in human cancer: A review, Cancer Res. **49** (1989), 4682–4689.
20. Buckley, I., Oncogenes and the nature of malignancy, Adv. in Cancer Res **50** (1988), 71–93.
21. Buick, R. N. and Pollak, M. N., Perspective on clonogenic tumor cells, stem cells and oncogenes, Cancer Res **44** (1984), 4909–4918.
22. Burck, K. B., Liu, E. T. and Larrick, J. W., "Oncogenes: An Introduction to the Concept of Cancer Genes," Springer-Verlag, New York, 1988.
23. Burmer, G. C. and Loeb, L. A., Mutations in the KRAS2 oncogene during progressive stages of human colon carcinoma, Proc. Natl. Acad. Sci. USA **86** (1989), 2403–2407.
24. Bush, R. S. and Hill, R. P., Biologic discussions augmenting radiation effects and model systems, Laryngoscope **85** (1975), 1119–1133.
25. Butel, J. S., SV40 large T antigen: Dual oncogene, Cancer Survey **5** (1986), 343–366.
26. Cavenee, W. K. et al., Expression of recessive alleles by chromosomal mechanisms in retinoblastoma, Nature **305** (1983), 719–784.
27. Cavenee, W. E. et al., Genetic origin of mutations predisposing to retinoblastoma, Science **228** (1985), 501–503.

28. Cerni, C. et al., Successive steps in the process of immortalization identified by transfer of separate bovine papillomavirus genes in rat fibroblasts, Proc. Natl. Acad. Sci. USA **86** (1989), 3266–3270.

29. Cerutti, P. A., Prooxidant states and tumor promotion, Science **227** (1985), 375–381.

30. Chandar, N., Lombardi, B. and Locker, J., c-myc gene amplification during hepato-carcinogenesis by a choline-devoid diet, Proc. Natl. Acad. Sci. USA **86** (1989), 2703–2707.

31. Chaum, E. et al., Cytogenetic analysis of retinoblastoma: evidence for multifocal origin and in vivo gene amplification, Cytogenet. Cell Genet. **38** (1984), 82–91.

32. Chu, K. C., "Multievent model of carcinogenesis: A mathematical model for cancer causation and prevention. In carcinogenesis:," A Comprehensive survey Volume 8, Raven Press, N.Y., 1985, pp. 411–421.

33. Compere, S. J. et al., The ras and myc oncogenes cooperate in tumor induction in many tissues when introduced into midgestation mouse embryos by retroviral vectors, Proc. Natl. Acad. Sci. USA **86** (1989), 2224–2228.

34. Corominas, M. et al., Oncogene activation in human benign tumors of the skin (keratoacanthomas): Is HRAS involved in differentiation as well as proliferation?, Proc. Natl. Acad. Sci. USA **86** (1989), 6372–6376.

35. Croce, C. M. and Klein, G., Chromosome translocations and human cancer, Scient. Amer. (March, 1985).

36. Croce, C. M., Role of chromosome translocations in human neoplasia, Cell **49** (1987), 155–156.

37. DeCaprio, J. A. et al., SV40 large tumor antigen forms a specific complex with the product of the retinoblastoma susceptibility gene, Cell **54** (1988), 275–283.

38. Depinho R. et al., Myc family of cellular oncogens, J. Cell. Bioch **33** (1987), 257–304.

39. Detto, G. P., Weinberg, R. A. and Ariza, A., Malignant transformation of mouse primary keratinocytes by HaSV and its modulation by surrounding normal cells,, Proc. Natl. Acad. Sci. **85** (1988), 6389–6393.

40. Doll, R., The age distribution of cancer: Implications for models of carcinogenesis, Roy. Statist. Society **(A)** **134** (1971), 133–166.

41. Dryja, T. P. et al., Homozygosity of chromosome 13 in retinoblastoma, The New England J. Medicine **310** (1984), 550–553.

42. Dyson, N. et al., The human papillomavirus 16E7 oncoprotein is able to bind to retinoblastoma gene product, Science **243** (1989), 934–937.

43. Erisman, M. D., Scott, J. K. and Astrin, S. M., Evidence that the familial adenomatous polyposis gene is involved in a subset of colon cancers with a complementable defect in c-myc regulation, Proc. Natl. Acad. Sci. USA **86** (1989), 4264–4268.

44. Escot, C. et al., Genetic alternation of the c-myc protooncogene (MYC) in human primary breast carcinomas, Proc. Natl. Acad. Sci. USA **83** (1986), 4834–4838.

45. Evans, H. J., Chromosomes and Cancer: From molecules to man—An overview, In "Chromosomes And Cancer," Academic Press, N.Y. 1983, pp 333–352.

46. Farber, E., The multistep nature of cancer development, Cancer Res. **44** (1984), 4217–4223.

47. Farber, E., Experimental induction of hepatocellular carcinoma as a paradigm for carcinogenesis, Clin. Physiol. Biochem **5** (1987), 152–159.

48. Farber, E., Rotstein, J. B. and Eriksson, L. C., Cancer development as a multistep process: Experimental results in animals, In "Mechanisms of Environmental Carcinogenesis, Vol II—Multistep Models of Carcinogenesis." Edited by J. C. Barrett, CRC Press, Boca Raton, 1986.

49. Farber, E. and Sarma, D. S. R., Hepatocarcinogenesis: A dynamic cellular perspective, Laboratory Investigation **56** (1987), 4–22.

50. Fasano, O. et al., Analysis of the transforming potential of the human H–ras gene by random mutagenesis, Proc. Natl. Acad. Sci. USA **81** (1984), 4008–4012.

51. Fearson, E. R., Volgestein, B. and Feinberg, A. P., Somatic deletion and duplication of genes on chromosome 11 in Wilm's tumors, Nature **309** (1984), 174–176.

52. Folkman, J. et al., Induction of angiogenesis during the transition from hyperplasia to neoplasia, Nature **339** (1989), 58–61.

53. Fong, C. T. et al., Loss of heterozygosity for the short arm of chromosome 1 in human neuroblastomas: Correlation with N-myc amplification., Proc. Natl. Acad. Sci. USA **86** (1989), 3753–3757.

54. Franza, B. R. et al., In vitro establishment is not a sufficient prerequisite for transformation by activated ras oncogenes, Cell **44** (1986), 409–418.

55. Friend, S. H. et al., A human DNA segment with properties of the gene that predisposes to retinoblastoma and osteosarcoma, Nature **323** (1986), 643–646.
56. Fung, J. K. T. et al., Structural evidence for the authenticity of the human retinoblastoma gene, Science **236** (1987), 1657–1660.
57. Glaichenhaus, N. et al., Cooperation between multiple oncogenes in rodent embryo fibroblasts: An experimental model of tumor progression, Advances in Cancer Research **45** (1985), 291–305.
58. George, D. L., Amplification of cellular proto–oncogenes in tumors and tumor cell lines, Cancer Survey **3** (1984), 499–513.
59. Green, A. R., Recessive mechanisms of malignancy, Br. J. Cancer **58**, 115–121.
60. Hanley, M. R. and Jackson, T., The ras gene: Transformer and transducer, Nature **328** (1987), 668–669.
61. Hansen, M. F. and Cavene, W. K., Genetics of cancer predisposition, Cancer Res. **47** (1987), 5518–5527.
62. Harris, H., The genetic analysis of malignancy, Cell Science Suppl. **4** (1986), 431–444.
63. Harrour, J. W. et al., Abnormalities in structure and expression of the human retinoblastoma gene in SCLC, Science **241** (1988), 353–357.
64. Hennings, H. et al., Malignant conversion of mouse skin tumors is increased by tumor initiators and unaffected by tumor promoters, Nature **304** (1983), 67–69.
65. Hennings, H. and Yuspa, S. H., Two–stage tumor promotion in mouse skin: An alternative interpretation, Jour. Nat. Cancer Inst. **74** (1986), 735–740.
66. Henry, I. et al., Tumor-specific loss of 11p15.5 in del11p13 Wilms tumor and in familial adrenocortical carcinoma, Proc. Natl. Acad. Sci USA **86** (1989), 3247–3251.
67. Horowitz, J. et al., Point mutational inactivation of the retinoblastoma antioncogene, Science **243** (1989), 937–940.
68. Hung, M.-C., Yan, D.-H. and Zhao, X., Amplification of the proto-neu oncogene facilitates oncogenic activation by a single point mutation, Proc. Natl. Acad. Sci. USA **86** (1989), 2545–2248.
69. Hurlin, P. J. et al., Morphological transformation, focus formation, and anchorage independence induced in diploid human fibroblasts by expression of a transfected H-ras oncogene, Cancer Res **47** (1987), 5752–5757.

70. Hurlin, P. J., Maher, V. M. and McCormick, J. J, Malignant transformation of human fibroblasts caused by expression of a transfected T24 HRAS oncogene, Proc. Natl. Acad. Sci. USA **86** (1989), 187–191.

71. Kalunaga, T., Requirement for cell replication in the fixation and expression of the transformed state in mouse cells treated with 4–nitroquinoline–1–oxide, Int. J. Cancer **14** (1974), 736–742.

72. Kennedy, A. R. and Little, J. B., Evidence indicating that the second step in X–ray induced transformation in vitro occurs during cellular proliferation, Radiat Res **99** (1984), 228–248.

73. Kerbel, R. S., Frost, P. and Liteplo, R. G., Genetic and epigenetic regulations of the metastatic phenotype, In "Biochemistry and Molecular Genetics of Metastasis," (eds. Liotta, L., Rabson, C. and Chambers, B.), Martinus Nijhoff/Dr. J. W. Junk Publ., Amsterdam, 1985.

74. Klawansky, S. and Fox, M. S., A growth rate distribution model for the age dependence of human cancer incidence: A proposed role for promotion in cancer of the lung and breast, Jour. Theor. Biol. **111** (1984), 531–587.

75. Klein, G., The approaching era of the tumor suppressor genes, Science **238** (1987), 1539–1545.

76. Klein, G. and Klein, E., Oncogene activation and tumor progression, Carcinogenesis **5** (1984), 429–436.

77. Klein, G. and Klein E., Evolution of tumors and the impact of molecular oncology, Nature **315** (1985), 190–195.

78. Knudson, A. G., Genetics and etiology of human cancer, Advances in Human Genetics **8** (1977), 1–66.

79. Knudson, A. G., Cancer genes in man, An International Colloquium, M. D. Anderson Tumor Hospital, Houston, TX, Nov 1981.

80. Knudson, A. G., Hereditary cancer, oncogenes and antioncogenes, Cancer Res **45** (1985), 1437–1443.

81. Koufos, A. et al., Loss of alleles at loci on human chromosome 11 during genesis of Wilm's tumor, Nature **309** (1984), 170–172.

82. Kovacs, G. et al., Consistent chromosome 3p deletion and loss of heterozygosity in renal cell carcinoma, Proc. Natl. Acad. Sci. USA **85** (1988), 1571–1575.

83. Kozumbo, W. J. and Cerutti, P. A., Antioxidants as antitumor promoters, In "Antimutagenesis and Anticarcinogenesis Mechanisms,"

p 491–508, Edited by D. M. Shankel, P. E. Hartman, T. Kada and A. Hollaender, Plenum Press, N.Y., 1986.

84. Lacal, J. C., Moscat, J. and Aaronson, S. A., Novel source of 1, 2-diacylglycerol elevated in cells transformed by Ha-ras oncogene, Nature **330** (1987), 269–272.

85. Land, H., Parada, L. F., and Weinberg, R. A., Tumorigenic conversion of primary embryo fibroblasts requires at least two cooperating oncogenes, Nature **304** (1983), 596–601.

86. Land, H., et al., Behavior of myc and ras oncogenes in transformation of rat embryo fibroblasts, Mol. Cell Biol **6** (1986), 1917–1925.

87. Langlois, R. G. et al., Evidence for increased in vivo mutation and somatic recombination in Bloom's syndrome, Proc. Natl. Acad. Sci. USA **86** (1989), 670–674.

88. Lee, W.-H, et al., The retinoblastoma susceptibility gene encodes a nuclear phosphoprotein associated with DNA binding activity, Nature **329** (1987), 642–645.

89. Lee, Y.-H. P., et al., Inactivation of the retinoblastoma susceptibility gene in human breast cancers, Science **241** (1988), 218–221.

90. Ling, V., Chambers, A. F., Harris, J. F. and Hill, R. P., Quantitative genetic analysis of tumor progression, Cancer and Metastasis Reviews **4** (1985), 173–194.

91. Liotta, L. A., Tumor invasion and metastases—Role of the extracellular matrix: Rhoads memorial award lecture, Cancer Research **46** (1986), 1–7.

92. Lowy, D. R. and Willumsen, B. M., The ras gene family, Cancer Survey **5** (1986), 275–289.

93. Mackillop, W. J. and Buick, R. N., Cellular heterogeneity in human ovarian carcinoma studied by density gradient fractionation, Stem cells **1** (1981), 355–366.

94. Mackillop, W. J., Ciampi, A. and Buck, R. N., A stem cell model of human tumor growth: Implications for tumor cell clonogenic assays, J. Nat. Cancer Inst **70** (1983), 9–16.

95. Mackillop, W. J., Stewart, S. S. and Buick, R. N., Density/volume analysis in the study of cellular heterogeneity in human ovarian carcinoma, Br. Jour Cancer **45** (1982), 812–820.

96. Marshall, C. J., Oncogenes, J. Cell Sci. Suppl. **4** (1986), 417–430.

97. Marshall, C. J. and Ridby, P. W. J. C., Viral and cellular genes involved in oncogenesis, Cancer Survey **3** (1984), 183–214.

98. Mathew, C. G. P. et al., Deletion of genes in chromosome 1 in endocrine neoplasia, Nature **328** (1987), 524–526.
99. Matsumura, T., Hayashi, M. and Konishi, R., Immortalization in culture of rat cells: A genealogic study, J. Nat. Cancer Inst. **74** (1985), 1223–1232.
100. Marx, J. L., The Yin and Yang of cell growth control, Science **232** (1986), 1093–1095.
101. Marx, J. L., Oncogene action probed, Science **237** (1987), 602–603.
102. Medina, D., The preneoplastic state in mouse mammary tumorigenesis, Carcinogenesis **9** (1988), 1113–1119.
103. Moolgavkar, S. H., Hormones and multistage carcinogenesis, Cancer Survey **3** (1986), 183–214.
104. Moolgavkar, S. H., Carcinogenesis modeling: From molecular biology to epidemiology, Ann. Rev. Public Health **7** (1986), 151–169.
105. Moolgavkar, S. H. and Venzon, D. J., Two event model for carcinogenesis: Incidence curves for childhood and adult cancer, Math. Biosciences **47** (1979), 55–77.
106. Moolgavkar, S. H. and Knudson, A. G., Mutation and cancer: A model for human carcinogenesis, Jour. Nat. Cancer Inst. **66** (1981), 1037–1052.
107. Nettesheim, P. and Barrett, J. C., In vitro transformation of rat tracheal epithelial cells as a model for the study of multistage carcinogenesis, In: "Carcinogenesis, Vol 9", 283–292, Edited by J. C. Barrett and R. W. Tennant, Raven Press, New York, 1985..
108. Nishizuka, Y., The molecular heterogeneity of protein kinase C and its implications for cellular regulation, Nature **334** (1988), 661–665.
109. Nowell, P. C., Mechanism of tumor progression, Cancer Research **46** (1986), 2203–2207.
110. Orkin, S. H. et al., Development of homozygosity for chromosome 11p markers in Wilm's tumors, Nature **309** (1984), 172–174.
111. Oshimura, M., Gilmer, T. M. and Barrett, J. C., Nonrandom loss of chromosome 15 in Syrian hamster tumors induced by v–Ha–ras plus v–myc oncogenes, Nature **316** (1985), 636–639.
112. Paterson, M. C. et al., Radiogenic neoplasia, cellular radiosensitivity and faulty DNA repair, In "Radiation Carcinogenesis: Epidemiology and Biological Significance," p 319–336, Raven Press, N.Y., 1984.
113. Pienta, K. J., Partin, A. W. and Coffey, D. S., Cancer as a disease of DNA organization and dynamic cell structure, Cancer Res. **49** (1989), 2525–2532.

114. Pitts, J. D., Kam, E. and Morgan, D., The role of junctional communication in cellular growth control and tumorigenesis, In: Modern Cell Biology Vol 7, eds, by Hertzberg, E. L. and Johnson, R. G. p 397–409, Alan R. Liss, Inc., N.Y. (1988).

115. Ponder, B., Gene losses in human tumors, Nature **335** (1988), 400–402.

116. Pryor, W. A., Cancer and free radicals, In "Antimutagenesis and Anticarcinogenesis Mechanisms," p 45–60, Edited by D. M. Shankel, P. E. Hartman, T. Kada and A. Hollaender, Plenum Press, New York, 1986.

117. Rowley, J. D., Implications of consistent chromosome rearrangements, In "Genes and Cancer," p 503–524, Alan R. Liss. Inc., New York, 1984.

118. Ruley, H. E. et al., Multistep transformation of an established cell line by the adenovirus Ela and T24 Ha–ras–1 genes, In "Cancer Cells 3, Growth Factors And Transformation," Edited by Feramisco, J. Ozanne, B. and Stiles, C., Cold Spring Harbor Lab, 1985.

119. Sager, R., Genetic suppression of tumor formation: A new frontier in cancer research, Cancer Research **46** (1986), 1573–1580.

120. Scherer, E. et al., Initiation–promotion–initiation. Induction of neoplastic foci within islands of precancerous liver cells in the rat, In "IARC Scientific Publication No. 56," Lyon, France, 1984.

121. Scott, R. E. et al., Biological mechanisms of stem cell carcinogenesis: A concept for multiple phases in the initiation and the role of differentiation control defects, "In Carcinogenesis, Vol 9," pp 67–79, Edited by J. C. Barrett and R. W. Tennant, Raven Press, New York, 1985.

122. Sinn, E. et al., Coexpression of MMTV/v-Ha-ras and MMTV/c-myc genes in transgenic mice: Synergistic action of oncogenes in vivo, Cell **49** (1987), 465–475.

123. Sirover, M. A. et al., Cellular and Molecular regulation of DNA repair in normal human cells and in hypermutable cells from cancer prone individuals, In "International Conference on Mechanisms of DNA Damage and Repair," June 2–7, 1985, edited by M. G. Simil, L. Grossman and A. C. Upton, Plenum Press, N.Y. 1986.

124. Sheer, D., Spurr, N. K. and Solomon, E., Gene mapping with special reference to human malignant diseases, Cancer Survey **3** (1984), 543–566.

125. Slaga, T. J., Overview of tumor promotion in animals, Env. Health Perspec. **50** (1983), 3–14.

126. Slaga, T. J., Cellular and molecular mechanisms of tumor promotion, Cancer Survey **2** (1983), 595–612.

127. Slaga, T. J., Mechanisms involved in multistage skin tumor carcinogenesis, In "Carcinogenesis, Vol 10," p 189–199, edited by E. Huberman and S. H. Barr, Raven Press, New York, 1985.

128. Spandidos, D. A. and Wilkie, N. M., Malignant transformation of early passage rodent cells by a single mutated human oncogene H–ras–1 from T 24 bladder carcinomaline, Nature **310** (1984), 469–475.

129. Stein, G. H., Namba, M. and Corsaro, C. M., Relationship of finite proliferative lifespan, senescence and quiescence in human cells, Jour. Cell. Physiol **122** (1985), 343–349.

130. Strauss, B. S., Cellular aspects of DNA repair, Advances in Cancer Research **45** (1985), 45–105.

131. Suarez, H. G. et al., Multiple activated oncogenes in human tumors, Oncogene Res. **1** (1987), 201–207.

132. Tan, W. Y., Some mixed models of carcinogenesis, Mathl. Comp. Modelling **10** (1988), 765–773.

133. Tan, W. Y. and Gastardo, M. T., On the assessment of effects of environmental agents on cancer tumor development by a two-stage model of carcinogenesis, Math Biosciences **74** (1985), 143–155.

134. Tan, W. Y., Chu, K. C. and Brown, C. C., The stochastic multi-event model for cancer: A stochastic multi-stage model with birth and death processes, National Cancer Inst./NIH,DCPC/BB report, Bethesda, MD. (1985).

135. Tan, W. Y. and Brown, C. C., A nonhomogeneous two-stage model of carcinogenesis, Math Modelling **9** (1987), 631–642.

136. Tan, W. Y. and Singh, K., A mixed model of carcinogneesis–with applications to retinoblastoma, Math. Biosciences **98** (1990), 201–211.

137. Tan, W. Y. and Chen, C., A stochastic model of carcinogenesis–Multiple pathways, Chapter 31 of Mathematical Population Dynamics, eds. O. Arino, D. E. Axelrod & M. Kimmel, Marcel Dekker Inc., 1990.

138. T'Ang, A., et al., Structural rearrangement of the retinoblastoma gene in human breast carcinoma, Science **242** (1988), 263–266.

139. Terranova, V. P., Hujanen, E. S. and Martin, G. R., Basement membrane and the invasive activity of metastatic tumor cells, Jour. Natl. Cancer Inst. **77** (1986), 311–316.

140. Thomassen, D. G. et al., Evidence for multiple steps in neoplastic transformation of normal and preneoplastic Syrian hamster embryo cells following transfection with Harvey murine sarcoma virus oncogene (v–Ha–ras), Cancer Research **45** (1985), 726–732.

141. Topal, M. D., DNA repair, oncogenes and carcinogenesis, Carcinogenesis **9** (1988), 691–696.

142. Trosko, J. E., Mechanisms of tumor promotion: Possible role of inhibited intercellular communication, Eur. J. Cancer **23** (1987), 599–601.

143. Trosko, J. E. and Chang, C. C., Role of intercellular communication in modifying the consequences of mutations in somatic cells, In "Antimutagenesis and Anticarcinogenesis Mechanisms," edited by D. M. Shankel, P. E. Hartman, T. Kada and A. Hollaender, p 439–458, Plenum Press, N.Y., 1986.

144. Trosko, J. E., Chang, C., and Medcalf, A., Mechanisms of tumor promotion: Potential role of intercellular communication, Cancer Invest. **1** (1983), 511–526.

145. Trosko, J. E. et al., Modulation of gap junction inter-cellular communication by tumor promoting chemicals, oncogenes, and growth factor during carcinogenesis, In: Modern Cell Biology **7**, eds. by Hertzberg, E. L. and Johnson, R. G., p435–448, Alan R. Liss, Inc., N.Y. 1988.

146. Tsunokawa, Y. et al., Integration of v–ras does not necessarily transform an immortalized murine cell line, Gann **75** (1984), 732–736.

147. Vogelstein, B., et al., Genetic alternations during colorectal-tumor development, N. Engl. J. Med. **319** (1988), 525–532.

148. Weinberg, R. A., The action of oncogenes in the cytoplasm and nucleus, Science **230** (1985), 770–776.

149. Weinberg, R. A., Finding the antioncogene, Scientific Amer. **259** (1988), 44–51.

150. Weinberg, R. A., The genetic origin of human cancer, Cancer **61** (1988), 1963–1968.

151. Weinberg, R. A., Oncogenes, antioncogenes and the molecular bases of multistep carcinogenesis, Cancer Research **49** (1989), 3713–3721.

152. Weichselbaum, R. R., Beckett, M. and Diamond, A., Some retinoblastomas, osteosarcomas, and soft-tissure sarcomas may share a common etiology, Proc. Natl. Acad. Sci. USA **85** (1988), 2106–2109.

153. Weissman, B. E. et al., Introduction of a normal human chromosome 11 into a Wilm's tumor cell line controls its tumorigenic expression, Science **236** (1987), 175–180.

154. Weston, K. and Bishop, J. M., Transcriptional activation by the v-myb oncogene and its cellular progenitor, c-myb, Cell **58** (1989), 85–93.

155. Weston, A. et al., Differential DNA sequence deletions from chromosomes 3, 11, 13 and 17 in squamous-cell carcinoma, large-cell carcinoma, and adenocarcinoma of the human lung, Proc. Natl. Acad. Sci. USA **86** (1989), 5099–5103.

156. Westway, D., Payne, G. and Varmus, H., Proviral deletions and oncogene base–substitutions in insertionally mutagenized c–myc alleles may contribute to the progression of avian bursal tumors, Proc. Natl. Acad. Sci. USA **81** (1984), 843–847.

157. Whittemore, A., Quantitative theory of oncogenesis, Advances in Cancer Research **27** (1978), 55–88.

158. Whittemore, A. and Keller, J. B., Quantitative theory of carcinogenesis, SIAM Review **20** (1978), 1–30.

159. Whyte, P. et al., Association between an oncogene and an anti-oncogene: The adenovirus ElA proteins bind to retinoblastoma gene product, Nature **334** (1988), 124–129.

160. Wolfman, A. and Macara, I. G., Elevated levels of diacylglycerol and decreased phorbol ester sensitivity in ras-transformed fibroblasts, Nature **325** (1987), 359–361.

161. Yamasaki, H., Role of gap junctional intercellular communication in malignant cell transformation, In: Modern Cell Biology, Vol 7, eds, by Hertzburg, E. L. and Johnson, R. G. p449–465, Alan R. Liss, Inc., N.Y., 1988.

162. Yuspa, S. H. and Poirier, M. C., Chemical carcinogenesis: From animal models to molecular models in one decade, Adv. in Cancer Res. **50** (1988), 25–70.

163. Zarbl, H. et al., Molecular assays for detection of ras oncogenes on human and animal tumors, In "Carcinogenesis Vol 9," p 1–16, edited by J. C. Barrett and R. W. Tennant, Raven Press, New York, 1985.

164. Zarbl, H. et al., Direct mutagenesis of Ha–ras–1 oncogenes by N–nitroso–N–methylurea during initiation of mammary carcinogenesis in rats, Nature **315** (1985), 382–385.

2

Some Basic Mathematical Tools and Stochastic Processes for Modeling Processes of Carcinogenesis

In order to develop stochastic models for the processes of carcinogenesis, I proceed in this chapter to develop some useful mathematical tools and present some stochastic processes which are useful in modeling processes of carcinogenesis. The results in this chapter will be used in later chapters to develop stochastic models of carcinogenesis under various biological conditions.

In Section (2.1), the incidence function and the probability generating function of the number of tumors are defined and some useful results are proved. To describe the stochastic growth of stem cells, intermediate cells and tumor cells, in Sections (2.2) and (2.3), three important stochastic birth-death processes-the nonhomogeneous Poisson processes, the non-homogeneous Feller-Arley birth-death processes and the stochastic logistic birth-death processes are discussed and some useful results are proved. In Section (2.4), we introduce the filtered Poisson processes and some of its properties. To conclude this chapter, in Section (2.5) we illustrate how to solve some first order partial differential equations and a linear Ricatti equation. These equations arise very often in modeling carcinogenesis stochastically as we shall see in later chapters.

2.1 THE INCIDENCE FUNCTION AND THE PROBABILITY GENERATING FUNCTION FOR THE NUMBER OF TUMORS

In modeling processes of carcinogenesis, an important summary quantity is the age-specific incidence function (or hazard function). To obtain the incidence function and the probability $P_j(t)$ of the number of tumors, an important mathematical tool is the probability generating function (PGF) of the number of tumors. In this section I give a brief introduction to the incidence function and the probability generating function of the number of tumors; I shall derive some results which are useful in developing stochastic models of carcinogenesis.

2.1.1 The Incidence Function of Cancer Tumors

Let T be the time that a cancer tumor develops for the first time given that at t_0 there are no cancer tumors. Let $f_T(t)$ be the probability density function (pdf) of T and $F_T(t)$ the cumulative distribution function (cdf) of T so that $f_T(t) = dF_T(t)/dt$. (Absolute continuous T.) Then $S_T(t) = 1 - F_T(t)$ is called the survival function of T and the incidence function $\lambda(t)$ of tumor onset at time t is defined by:

$$\lambda(t) = \lim_{\Delta t \to 0} Pr\{t \le T \le t + \Delta t | T \ge t\}/\Delta t$$
$$= f_T(t)/S_T(t) = f_T(t)/[1 - F_T(t)]$$
$$= -d \log S_T(t)/dt, \tag{2.1.1}$$

From (2.1.1), obviously

$$S_T(t) = 1 - F_T(t) = \exp\{-\int_{t_0}^{t} \lambda(x)dx\} \tag{2.1.2}$$

where t_0 is the time of birth of the individual.

It follows that

$$f_T(t) = -dS_T(t)/dt = \lambda(t)\exp\{-\int_{t_0}^{t} \lambda(x)dx\},$$
$$t \ge t_0, \tag{2.1.3}$$

Note that $\lim_{t \to \infty} F_T(t) = 1$ implies $\int_{t_0}^{\infty} \lambda(x)dx = \infty$. Formulas (2.1.1)–(2.1.3) imply that given $\lambda(t)$ one may derive $f_T(t)$ while given $f_T(t)$ (or $S_T(t)$ or $F_T(t)$) one may derive $\lambda(t)$.

EXAMPLE (2.1)   THE ARMITAGE-DOLL MODEL   In this model, $\lambda(t) = c(t - t_0)^{k-1}$, where c $(c > 0)$, and k $(k \geq 1)$ are constants. Hence,

$$\int_{t_0}^{t} \lambda(x)dx = (c/k) \int_{t_0}^{t} k(x - t_0)^{k-1}dx$$

$$= (c/k) \int_{t_0}^{t} d(x - t_0)^k = (c/k)[(t - t_0)^k];$$

and $f_T(t) = c(t - t_0)^{k-1}\exp\{-(c/k)[(t - t_0)^k]\}$, $t \geq t_0$.

EXAMPLE (2.2)   Let

$$\lambda(t) = \exp\{-\int_{t_0}^{t} \epsilon(x)dx\}\alpha(t) = u(t_0, t)\alpha(t).$$

Then

$$\int_{t_0}^{t} \lambda(x)dx = \int_{t_0}^{t} u(t_0, x)\alpha(x)dx;$$

hence,

$$f_T(t) = u(t_0, t)\alpha(t)\exp\{-\int_{t_0}^{t} u(t_0, x)\alpha(x)dx\} \text{ for } t \geq t_0.$$

REMARK   Example (2.2) is the incidence function of a one-stage model of carcinogenesis; see Example (2.14) in Section (2.3).

EXAMPLE (2.3)   Let the incidence function $\lambda(t)$ be given by:

$$\lambda(t) = \alpha + \beta \exp(\gamma t),$$

where α, β and γ are real constants, where $\alpha > 0$ and $\beta > 0$. (This incidence function was referred to as Makeham's law and widely used by actuaries.) Then, by taking the derivative, it is easily seen that $\lambda(t)$ is strictly increasing if $\gamma\beta > 0$, constant if $\gamma\beta = 0$ and strictly decreasing if $\gamma\beta < 0$.

$$\text{If } \gamma \neq 0, \quad \int_{0}^{t} \lambda(x)dx = \alpha t + \beta \int_{0}^{t} \exp(\gamma x)dx$$

$$= \alpha t + (\beta/\gamma)[\exp(\gamma t) - 1]$$

which goes to ∞ as $t \to \infty$. (Note $\lambda(t) > 0$ for all $t \geqslant 0$) Hence,

$$S(t) = \exp\left\{-\int_0^t \lambda(x)dx\right\} = \exp\left\{-[\alpha t + (\beta/\gamma)(e^{\gamma t} - 1)]\right\},$$

and

$$f_T(t) = [\alpha + \beta \exp(\gamma t)] \exp\{-[\alpha t + (\beta/\gamma)(e^{\gamma t} - 1)]\}, t \geq 0.$$

EXAMPLE (2.4) THE COX HAZARD FUNCTION While the hazard function (incidence function) is specified by the carcinogenesis model, to identify etiological agents of cancer and to develop statistical procedures for risk assessment of environmental agents, in the past statisticians would assume some specific hazard functions without due regard to the biological process of carcinogenesis. A commonly used hazard function is the Cox's model (2) to relate environmental factors $\underset{\sim}{Z}$ (called covariates) to the hazard function. Cox's model assumes that

$$\lambda(t) = \lambda_0(t)R(\underset{\sim}{Z}, \underset{\sim}{\beta}), \tag{2.1.4}$$

with $\lambda_0(t) > 0$ and $R(\underset{\sim}{Z}, \beta) > 0$, where $\lambda_0(t)$ has been referred to as a baseline hazard function while $R(\underset{\sim}{Z}, \beta)$ the risk function which relates the covariates $\underset{\sim}{Z}$ with the hazard function, $\underset{\sim}{\beta}$ being the vector of unknown parameters.

In this case, if $R(\underset{\sim}{Z}, \underset{\sim}{\beta})$ is independent of t,

$$S(t) = 1 - \exp\left\{-R(\underset{\sim}{Z}, \underset{\sim}{\beta}) \int_{t_0}^t \lambda_0(x)dx\right\},$$

and hence,

$$f_T(t) = \lambda_0(t)R(\underset{\sim}{Z}, \underset{\sim}{\beta}) \exp\left\{-R(\underset{\sim}{Z}, \underset{\sim}{\beta}) \int_{t_0}^t \lambda_0(x)dx\right\}, t \geqslant t_0.$$

The function

$$\int_{t_0}^t \lambda_0(x)dx$$

is referred to as the baseline integrated hazard function.

2.1.2 The Probability Generating Function of the Number of Cancer Tumors

Let $P_j(t)$ be the probability of having j cancer tumors at time t given that there are no tumors at time $t_0(t \geq t_0)$. Then the probability generating function (PGF) Q(s,t) of $\{P_j(t), j = 0, 1, 2, \ldots\}$ is defined by:

$$Q(s,t) = \sum_{j=0}^{\infty} s^j P_j(t). \tag{2.1.5}$$

As we shall see in Chapters 3–6, for developing stochastic models for carcinogenesis the basic approach is to obtain Q(s, t).

EXAMPLE (2.5)   PGF OF A POISSON PROCESS

Let $P_j(t) = e^{-\lambda(t)}(\lambda(t))^j/j!, j = 0, 1, 2, \ldots,$ where $\lambda(t) > 0$. Then,

$$Q(s,t) = \sum_{j=0}^{\infty} s^j P_j(t) = \exp(-\lambda(t)) \sum_{j=0}^{\infty} [s\lambda(t)]^j/j!$$
$$= \exp\{-\lambda(t) + s\lambda(t)\} = \exp[(s-1)\lambda(t)].$$

EXAMPLE (2.6)   PGF OF A FILTERED POISSON PROCESS   Let $N(t), t \geq t_0$, be a Poisson variate with parameter $(t - t_0)\lambda(t)$ $(\lambda(t) > 0)$ and let the $U_j's$ be independently and uniformly distributed over $[t_0, t]$. Let $Z(t), t \geq t_0$, be defined by

$$Z(t) = \sum_{j=1}^{N(t)} \omega(t, U_j, Y_j), \tag{2.1.6}$$

where $\omega(t, U, Y)$ is a real-valued function of (t, U, Y) satisfying $\omega(t, U, Y) = 0$ if $t < U$ and where the $Y_j's$ are independently and identically distributed random variables which are further assumed to be independently distributed of the $U_j's$ and $N(t)$.

The stochastic process $Z(t)$ as defined above has been termed by Parzen (11) as a filtered Poisson process. Let $Q(s,t)$ be the PGF of $Z(t)$. I now proceed to show that $Q(s,t)$ is given by:

$$Q(s,t) = \exp\{\lambda(t) \int_{t_0}^{t} [\phi(s;u,t) - 1]du\}$$

where $\phi(s;u,t)$ is the conditional PGF of $\omega(t, U, Y)$ given $U = u$.

To obtain $Q(s,t)$, observe first that since U is uniformly distributed over $[t_0, t]$, the PGF of $\omega(t, U, Y)$ is

$$\zeta(t_0, t) = \int_{t_0}^{t} \phi(s; u, t) du / (t - t_0).$$

It follows that the conditional PGF of $Z(t)$ given $N(t) = j$ is $[\zeta(t_0, t)]^j$. Since $N(t)$ is Poisson with parameter $(t - t_0)\lambda(t)$, by taking expectation of $[\zeta(t_0, t)]^{N(t)}$ we obtain:

$$Q(s,t) = \exp\{-\lambda(t)(t - t_0) + \lambda(t)(t - t_0)\zeta(t_0, t)\}$$

$$= \exp\{\lambda(t) \int_{t_0}^{t} [\phi(s; u, t) - 1] du\}.$$

Parzen (11) showed that the process $Z(t)$ as represented above can be used to describe many engineering processes. Specific examples include the number of claims in force on a workman's compensation insurance policy and the number of pulses locking a paralyzable counter. I shall show in Section (2.4) and in Chapter 3 that many processes of carcinogenesis are in fact filtered Poisson processes; for specific examples from biomedical research, see Tan (16).

Application of PGF to Derive Incidence Functions

Given the PGF $Q(s,t)$ of the number of tumors at time t, one may not only compute the probability of having j tumors at time t, but also can use the following theorem to derive the incidence function $\lambda(t)$ of tumors.

THEOREM (2.1.1)   If cancer tumors do not regress, then

$$\lambda(t) = -Q'(0,t)/Q(0,t), \tag{2.1.7}$$

where $Q'(0,t) = dQ(0,t)/dt$

Proof.   If cancer tumors do not regress, then $Q(0,t)$ is the probability that tumors develop after time t. It follows that $[Q(0,t) - Q(0, t + \Delta t)]/Q(0,t)$ is the conditional probability that a tumor develops during $[t, t + \Delta t]$ given that there are no tumors at or before time t. Hence,

$$\lambda(t) = \lim_{\Delta t \to 0} \frac{Q(0,t) - Q(0, t + \Delta t)}{(\Delta t) Q(0,t)} = -Q'(0,t)/Q(0,t).$$

In Chapters 3–6, I shall use Theorem (2.1.1) to derive incidence functions of cancer tumors for various carcinogenesis models.

EXAMPLE (2.7) Let $Q(s,t) = \exp[(s-1)\int_{t_0}^{t} u(t_0,x)\alpha(x)dx], t \geqslant 0$. Then

$$Q'(0,t) = -\exp\{-\int_{t_0}^{t} u(t_0,x)\alpha(x)dx\}u(t_0,t)\alpha(t)$$

$$= -Q(0,t)u(t_0,t)\alpha(t).$$

Hence·

$$\lambda(t) = -Q'(0,t)/Q(0,t) = u(t_0,t)\alpha(t).$$

As we shall see, this incidence function is that of a one-stage model of carcinogenesis; see example (2.14) in Section (2.3).

EXAMPLE (2.8) Write $Q(s,t)$ as $Q(s,t) = Q(s;t_0,t)$ and let

$$Q(s;t_0,t) = \exp\{\int_{t_0}^{t} X(u)\alpha(u)[\psi(s;u,t) - 1]du\}, \text{ where } \psi(s;u,u) = 1.$$

Then

$$Q'(0;t_0,t) = Q(0;t_0,t)\{X(t)\alpha(t)[\psi(0;t,t) - 1]$$

$$+ \int_{t_0}^{t} X(u)\alpha(u)[\frac{\partial}{\partial t}\psi(0;u,t)]du\}$$

$$= -Q(0;t_0,t)\int_{t_0}^{t} X(u)\alpha(u)\mu(u,t)du,$$

where $\mu(u,t) = -\frac{\partial}{\partial t}\psi(0;u,t)$.

Hence

$$\lambda(t) = -Q'(0;t_0,t)/Q(0;t_0,t) = \int_{t_0}^{t} X(u)\alpha(u)\mu(u,t)du$$

As we shall see in Chapter Three, example (2.8) is related to the two-stage models of carcinogenesis.

A General Result for Computing Probabilities by Using PGF

Given the PGF $Q(s,t)$ of $\{P_j(t), j = 0,1,2,...\}$, theoretically one may obtain $P_j(t)$ from $Q(s,t)$ by the relationship:

$$(j!)P_j(t) = \{\partial^j Q(s,t)/\partial s^j\}_{s=0}, \tag{2.1.8}$$

As we shall see, for most of the carcinogenesis models, it is extremely difficult to obtain an explicit and closed form of $P_j(t)$ from (2.1.8). The following theorem (Theorem (2.1.2)) provides a procedure to compute $P_j(t)$ iteratively from $Q(s,t)$. The results in Theorem (2.1.2) are important for developing carcinogenesis models, since as we shall see in Chapters 3 and 4, for most of the carcinogenesis models it is easy to compute the $q_j(t)$ functions as defined in the theorem.

THEOREM (2.1.2)   Writing $Q(s,t)$ as

$$Q(s,t) = \exp\left\{-q_0(t) + \sum_{j=1}^{\infty} s^j q_j(t)\right\},$$

then, $P_j(t)$ can be computed iteratively by the following procedure:

$$P_0(t) = \exp\left\{-q_0(t)\right\}, \quad \text{and for} \quad j \geq 1,$$

$$P_j(t) = P_0(t)q_j(t) + \sum_{u=1}^{j-1} q_{j-u}(t)P_u(t)[(j-u)/j], \tag{2.1.9}$$

$$\text{where} \quad \sum_{u=1}^{0} \quad \text{is defined as zero.}$$

Proof. Obviously $P_0(t) = \exp[-q_0(t)]$. To prove the general result, let $M(s,t) = \log Q(s,t) = -q_0(t) + \sum_{j=1}^{\infty} s^j q_j(t)$ and note that $Q(s,t) = \sum_{j=0}^{\infty} s^j P_j(t)$. Then,

$$\sum_{j=1}^{\infty} j s^{j-1} q_j(t) = [\partial M(s,t)/\partial s] = [\partial Q(s,t)/\partial s]/Q(s,t).$$

Or, with $a_j(t) = (j+1)q_{j+1}(t), j = 0, 1, 2, \ldots,$

$$\left[\sum_{j=0}^{\infty} s^j a_j(t)\right]\left[\sum_{j=0}^{\infty} s^j P_j(t)\right] = \sum_{j=0}^{\infty} (j+1)s^j P_{j+1}(t).$$

By Cauchy product, the left side can be expressed as

$$\sum_{j=0}^{\infty} s^j C_j(t)$$

where

$$C_j(t) = a_0(t)P_j(t) + a_1(t)P_{j-1}(t) + \ldots + a_j(t)P_0(t)$$

$$= (j+1)q_{j+1}(t)P_0(t) + \sum_{m=1}^{j} m q_m(t)P_{j+1-m}(t)$$

$$= (j+1)q_{j+1}(t)P_0(t) + \sum_{u=1}^{j} (j+1-u)q_{j+1-u}P_u(t).$$

Comparing coefficients of both sides of the above formula, we have, for $j = 0, 1, \ldots$:

$$(j+1)P_{j+1}(t) = (j+1)q_{j+1}(t)P_0(t) + \sum_{u=1}^{j}(j+1-u)q_{j+1-u}(t)P_u(t).$$

Or

$$P_{j+1}(t) = q_{j+1}(t)P_0(t) + \sum_{u=1}^{j} q_{j+1-u}(t)P_u(t)[\frac{j+1-u}{j+1}], j = 0, 1, 2 \ldots.$$

The results in Theorem (2.1.2) were first given by Tan (14) and had been used extensively by Tan (14, 15, 16), Tan and Gastardo (18) and Tan and Brown (19). The results in Theorem (2.1.2) are important since as shown by Tan and Tou (22), as long as $P_j(t)$ is computable, one may readily develop maximum likelihood procedures for estimating unknown genetic parameters which in turn would lead to the computation of scoring statistics for testing statistical hypotheses concerning unknown parameters.

EXAMPLE (2.9) Let $Y(t), t \geq 0$, be a stochastic process with state space $S = \{0, 1, 2, \ldots\}$. Suppose that the PGF of $Y(t)$ is

$$\phi(s,t) = \exp\{\lambda \int_0^t [f(s,x) - 1]\exp[(b_1 - d_1)(t-x)]dx\},$$

where

$$f(s,t) = \{(s-1)d_2 + (d_2 - sb_2)\exp[-(b_2 - d_2)t]\}$$
$$\times \{(s-1)b_2 + (d_2 - sb_2)\exp[-(b_2 - d_2)t]\}^{-1}.$$

Put $\epsilon_i = b_i - d_i, i = 1, 2,$ and write $\phi(s,t)$ as $\phi(s,t) = \exp[-q_0(t) + \sum_{j=1}^{\infty} s^j q_j(t)].$
Then,

$$q_0(t) = \lambda\epsilon_2 \int_0^t \exp[\epsilon_1(t-x)][b_2 - d_2 \exp(-\epsilon_2 x)]^{-1} dx$$

and for $j = 1, 2, \dots,$

$$q_j(t) = \lambda\epsilon_2^2 \int_0^t \exp(-\epsilon_2 x)[b_2 - d_2 \exp(-\epsilon_2 x)]^{-2}[h(x)]^{j-1} dx,$$

where

$$h(x) = b_2[1 - \exp(-\epsilon_2 x)]/[b_2 - d_2\exp(-\epsilon_2 x)].$$

Let $P_j(t)$ satisfy the relationship $(j!)P_j(t) = \{d^j \phi(s,t)/ds^j\}_{s=0}$. Then, by Theorem (2.1.2), one may compute $P_j(t)$ by:

$$P_0(t) = \exp[-q_0(t)] \quad \text{and for } j = 1, 2, \dots,$$

$$P_j(t) = P_0(t)q_j(t) + \sum_{u=1}^{j-1} q_{j-u}(t)P_u(t)[(j-u)/j].$$

For example, if $\lambda = 0.3, b_1 = b_2 = 0.04, d_1 = d_2 = 0.004$ and $t = 12$, the $P_j(t)$'s are given in Table (2.1) as:

The PGF of Example (2.9) is the PGF for the number of mutants in a large population of bacteria or cells which are subjected to stochastic birth and death and mutation from nonmutants to mutants. It was

Table (2.1) $P_j(t)$ For $\lambda = 0.3$, $b_1 = b_2 = 0.04$, $d_1 = d_2 = 0.004$ and $t = 12$.

j	0	1	2	3	4	5	6	7
$P_j(t)$	.0121	.0432	.0850,	.1214	.1406	.1404	.1251	.1019
j	8	9	10	11	12	13	14	15
$P_j(t)$	.0772	.0551	.0373	.0343	.0152	.0092	.0054	.0031
j	16	17	18	19	20	21		
$P_j(t)$	.0017	.0009	.0005	.0003	.0001	.0001		

first given in Tan (14) and was applied by Tan (15) to develop stochastic models for the CHO/HGPRT bioassay for testing mutagenicity of chemicals.

EXAMPLE (2.10) Let $Y(t), t \geq 0$ be a stochastic process with state space $S = \{0, 1, 2, ...\}$ and PGF $Q(s,t)$ given by:

$$Q(s,t) = \exp\{\lambda \int_0^t N(x)[\phi(s, t-x) - 1]dx\},$$

where

$$N(x) = \exp(\alpha x)/[1 - (N_0/M) + (N_0/M)\exp(\alpha x)]$$

and where

$$\phi(s,t) = \{\beta_1(\beta_2 - 1) + \beta_2(1 - \beta_1)\exp[-b(\beta_2 - \beta_1)t]\}$$
$$\times \{(\beta_2 - 1) + (1 - \beta_1)\exp[-b(\beta_2 - \beta_1)t]\}^{-1},$$

$b > 0$ and $\beta_2 > \beta_1 \geq 0$ being fixed real numbers.
Let

$$\omega_j(t) = \{d^j \phi(s,t)/ds^j\}_{s=0}/(j!)$$

and

$$q_j(t) = (-1)^{\delta_{0j}}\lambda \int_0^t N(x)[\omega_j(t - x) - \delta_{j0}]dx.$$

Then, $Q(s,t) = \exp[-q_0(t) + \sum_{j=1}^{\infty} s^j q_j(t)]$. Hence, if $P_j(t) = \{d^j Q(s,t)/ds^j\}_{s=0}/(j!)$, then, by Theorem (2.1.2), $P_0(t) = \exp[-q_0(t)]$ and for $j = 1, 2, ...,$

$$P_j(t) = P_0(t)q_j(t) + \sum_{u=1}^{j-1} q_{j-u}(t)P_u(t)[(j-u)/j].$$

The PGF of Example (2.10) is that of a homogeneous two-stage model of carcinogenesis proposed by Moolgavkar and Knudson (9). Using Theorem (2.1.2), Tan and Gastardo (18) have computed the probability of the number of tumors for a model extending that of Moolgavkar and Knudson (9).

2.2 SOME GENERAL THEORIES OF STOCHASTIC BIRTH-DEATH PROCESSES USEFUL FOR MODELING PROCESSES OF CARCINOGENESIS

To develop stochastic models of carcinogenesis, it is essential to model growth of stem cells, intermediate cells as well as cancer tumor cells. For carcinogenesis models, the three common growth curves are the exponential growth, the logistic growth and the Gompertz growth curves. It has been shown that growth of breast stem cells is best described by logistic growth (7, 9) while growth of stem cells for the colon and the lung is best described by Gompertz growth. Laird (4, 5) and Simpson-Herren and Lloyd (12) have also demonstrated that growth of many cancer tumor cells is best fitted by Gompertz growth curves.

In the past, growth patterns of stem cells and tumor cells are described mainly by deterministic approaches. (See reference 3, Chapter 2.) In this book I shall develop stochastic growth models for these growth patterns. I find it necessary and important to develop stochastic growth models for the following reasons:

   i. Carcinogenesis is a random process. Hence it is more realistic to work with stochastic models than with deterministic models.

   ii. In some cases, deterministic models appear to be special cases of stochastic models if one restricts oneself to working with expected values.

  iii. One may obtain more information by stochastic models than by deterministic models.

2.2.1 Some General Stochastic Birth and Death Processes for Modeling Processes Of Carcinogenesis

Let $X(t), t \geq t_0$, be a Markov stochastic process with state space $S = \{0, 1, 2, \ldots\}$. (The sample space of $X(t)$ is called the state space of $X(t)$.) $X(t)$ is Markov if and only if for all $n \geq 1$ and for all $t_1 \leq t_2 \leq \ldots \leq t_n \leq t$,

$$Pr\{X(t) = j | X(t_1) = i_1, \ldots, X(t_n) = i_n\} = Pr\{X(t) = j | X(t_n) = i_n\}.$$

DEFINITION (2.2.1) $X(t)$ given above is called a general birth-death process with birth rate $b_j(t)$ and death rate $d_j(t)$ if and only if the following conditions are satisfied:

i. $Pr\{X(t + \Delta t) = j + 1 | X(t) = j\} = b_j(t)\Delta t + o(\Delta t)$, where

$$\lim_{\Delta t \to 0} o(\Delta t)/\Delta t = 0;$$

ii. $Pr\{X(t + \Delta t) = j - 1 | X(t) = j\} = d_j(t)\Delta t + o(\Delta t);$
iii. $Pr\{X(t + \Delta t) = k | X(t) = j\} = o(\Delta t)$ if $|k - j| \geqslant 2;$

In what follows, I shall call the above process $X(t), t \geqslant t_0$, a general density-dependent birth-death process with birth rate $b_j(t)$ and death rate $d_j(t)$.

Some Special Cases:

i. If $b_j(t) = b(t)$ and $d_j(t) = 0$ for all j and t, then $X(t)$ is a non-homogeneous Poisson process; if further $b(t) = b, X(t)$ is then a homogeneous Poisson process.
ii. If $b_j(t) = jb$ and $d_j(t) = jd, X(t)$ is a homogeneous birth-death process with birth rate b and death rate d. (In the past this process was referred to as a Feller-Arley process.) This is the process used by Moolgavkar and Venzon (8) and Moolgavkar and Knudson (9) for cell proliferation of initiated stem cells (or intermediate cells) in a homogeneous two-stage model of carcinogenesis.
iii. If $b_j(t) = jb(t)$ and $d_j(t) = jd(t)$, then $X(t)$ is a nonhomogeneous birth-death process with birth rate $b(t)$ and death rate $d(t)$. (This is the nonhomogeneous generalization of the Feller-Arley process.) Tan and Brown (19) applied the nonhomogeneous birth-death process to describe stochastic growth of initiated cells in a nonhomogeneous two-stage model of carcinogenesis.
    If $b_j(t) = jb(t)$ and $d_j(t) = jd(t)$ and if further $\gamma(t) = b(t) - d(t) = (b - d)\exp(-\lambda t)$ for some $\lambda > 0$, then $X(t)$ is a stochastic Gompertz birth-death process (17). It follows that stochastic Gompertz growth is a special case of the nonhomogeneous Feller-Arley process.
iv. If the sample space of S is finite, e.g. $S = \{0, 1, 2, ..., M\}$, then $X(t)$ is called a finite birth-death process. If $S = \{0, 1, 2, ..., M\}$ and if $b_j(t) = jb(t)[1 - (j/M)]$ and $d_j(t) = jd(t)[1 - (j/M)]$, then $X(t)$ is a nonhomogeneous stochastic logistic birth-death process (21).

EXAMPLE (2.11) The growth of bacteria in a container usually follows a logistic growth pattern because of the limitation of the size of the container. Let $X(t)$ be the number of bacteria at time t and let M be the maximum size. If the birth rate is b and the death rate is d, then $X(t)$ follows a homogeneous logistic birth-death process. Given $X(0) = N_0$, the expected value of $X(t)$ is closely approximated by:

$$E[X(t)|X(0) = N_0]$$
$$\cong N_0 \exp[(b-d)t]\{1 - (N_0/M) + (N_0/M)\exp[(b-d)t]\}^{-1}.$$

(See reference 21).
Logistic growth has also been used successively by Tan (15) to describe the growth of ovary cells of a Chinese hamster in the CHO/HGPRT bioassay for testing mutagenicity of chemicals. Moolgavkar and Knudson (9) showed that the growth of breast cells was best described by logistic growth in studying breast cancers of women.

EXAMPLE (2.12) GENETIC MODEL OF MORAN A model proposed by Moran (10) concerns two alleles A_1 and A_2 in a haploid population of fixed population size M together with the following basic assumptions:

i. During $[t, t + \Delta t]$, the probability of having more than one death is $o(\Delta t)$,
ii. An A_j individual has life time distribution $\lambda_j \exp(-\lambda_j t), t \geqslant 0$, where $\lambda_j > 0$ is independent of t ($\lambda_1 \neq \lambda_2$ with selection and $\lambda_1 = \lambda_2 = \lambda$ if there is no selection).
iii. Whenever a death occurs during $[t, t + \Delta t]$, it is replaced immediately by an individual which is A_1 or A_2 with respective probabilities $p_t^* = \frac{1}{M}X(t)(1-\alpha_1) + (1 - \frac{1}{M}X(t))\alpha_2$ and $q_t^* = 1 - p_t^*$, where α_1 and α_2 are the mutation rates from A_1 to A_2 and from A_2 to A_1 respectively, and where $X(t)$ is the number of A_1 alleles at time $t, t \geqslant 0$.

I now proceed to show that $X(t)$ is in fact a finite homogeneous birth-death process with state space $S = \{0, \ldots, M\}$ and with birth rate b_j and death rate d_j being given by

$$b_j = \lambda_2 p_j(M-j) \text{ and } d_j = \lambda_1 j q_j, \text{ where}$$

$$p_j = \frac{1}{M}j(1-\alpha_1) + (1 - \frac{1}{M}j)\alpha_2 \text{ and } q_j = 1 - p_j, j = 0, 1, 2 \ldots M.$$

Let now T_j be the time that an A_j individual survives until $T_j, j = 1, 2$. Then, by (ii) given above, the probability density of T_j is

$$f_j(t) = \lambda_j \exp(-\lambda_j t), t \geqslant 0 \quad \text{so that}$$

$$Pr\{T_j \geqslant t\} = \int_t^\infty \lambda_j \exp(-\lambda_j z) dz = \exp(-\lambda_j t);$$

hence $Pr\{An \; A_j \text{ individual dies during } [t, t + \Delta t]| \text{ an } A_j \text{ at time } t\}$

$$= 1 - Pr\{T_j \geqslant t + \Delta t | T_j \geqslant t\}$$

$$= 1 - Pr\{T_j \geqslant t + \Delta t\}/Pr\{T_j \geqslant t\}$$

$$= 1 - \exp[-(t + \Delta t)\lambda_j]/\exp(-\lambda_j t) = 1 - \exp(-\lambda_j \Delta t)$$

$$= \lambda_j \Delta t + o(\Delta t).$$

It follows that $Pr\{An \; A_1 \text{ individual dies during } [t, t + \Delta t]|X(t) = k\} = Pr\{An \; A_1 \text{ individual dies during } [t, t + \Delta t]| \text{ there are } k \; A_1 \text{ individuals at time } t\} = k\lambda_1 \Delta t + o(\Delta t)$ and $Pr\{An \; A_2 \text{ individuals dies during } [t, t + \Delta t]|X(t) = k\} = Pr\{An \; A_2 \text{ individual dies during } [t, t + \Delta t]| \text{ there are } M - kA_2 \text{ individuals at time } t\} = (M - k)\lambda_2 \Delta t + o(\Delta t)$. Thus, $P_{j,j+1}(\Delta t) = Pr\{X(t + \Delta t) = j + 1|X(t) = j\} = Pr\{An \; A_2 \text{ individual dies and is replaced by an } A_1 \text{ individual during } [t, t + \Delta t]|X(t) = j\} = (M - j)\lambda_2 p_j \Delta t + o(\Delta t); P_{j,j-1}(\Delta t) = Pr\{X(t + \Delta t) = j - 1|X(t) = j\} = Pr\{An \; A_1 \text{ individual dies and is replaced by an } A_2 \text{ individual during } [t, t + \Delta t]|X(t) = j\} = j\lambda_1 q_j \Delta t + o(\Delta t);$ and $P_{j,k}(\Delta t) = Pr\{X(t + \Delta t) = k|X(t) = j\} = o(\Delta t)$ if $|k - j| \geqslant 2$ by (i).

This shows that Moran's genetic model is in fact a finite homogeneous birth-death process with $b_j = (M - j)\lambda_2 p_j$ and $d_j = j\lambda_j q_j$ and state space $S = \{0, 1, 2, \ldots M\}$.

EXAMPLE (2.13) THE ONE-STAGE MODEL OF CARCINOGEN-ESIS In modeling processes of carcinogenesis involving animals or human beings, although multistage is a general rule, there are not very frequent occasions in which one stage is sufficient to induce malignant conversion of normal stem cells. (For specific examples, see Chapter One.) In Example (2.14), it will be shown that under some general conditions, the number $Y(t)$ of tumors at time t is a Poisson process.

2.2.2 The Kolmogorov Forward Equations and the PGF of Birth-Death Processes

Let $X(t), t \geq t_0$ be a general birth-death process with birth rate $b_j(t)$ and death rate $d_j(t), j = 0, 1, 2, \ldots$, as defined above. Let $P_j(t) = Pr\{X(t) = j | X(t_0) = 1\}$ for $t \geq t_0$. Then $P_j(t_0) = \delta_{1j}$, where δ_{ij} is the Kronecker's δ defined by $\delta_{ii} = 1$ and $\delta_{ij} = 0$ for $i \neq j$.

Given the birth and death rates, an important problem is to compute $P_j(t)$. The Kolmogorov forward equations provide mathematical equations for computing $P_j(t)$. These equations are given in the following Theorem.

THEOREM (2.2.1) Putting $b_{-1}(t) = P_{(-1)}(t) = 0$, then the $P_j(t)'s$ satisfy the following system of equations:

$$\frac{d}{dt} P_j(t) = -P_j(t)[b_j(t) + d_j(t)] + P_{j-1}(t)b_{j-1}(t)$$
$$+ P_{j+1}(t)d_{j+1}(t), \; P_j(t_0) = \delta_{1j}, j = 0, 1, \ldots \quad (2.2.1)$$

Equations (2.2.1) are called the Kolmogorov forward equations.

Proof. Partitioning the time interval $[t_0, t + \Delta t]$ by $[t_0, t]$ and $(t, t + \Delta t]$, then, by Chapman-Kolmogorov equation,

$$P_j(t + \Delta t) = \sum_{k=0}^{\infty} P_k(t) Pr[X(t + \Delta t) = j | X(t) = k]$$
$$= P_{j-1}(t)b_{j-1}(t)\Delta t + P_{j+1}(t)d_{j+1}(t)\Delta t$$
$$+ P_j(t)\{1 - [b_j(t) + d_j(t)]\Delta t\} + o(\Delta t).$$

Subtracting $P_j(t)$ from both sides and dividing by Δt, one has:

$$\{P_j(t + \Delta t) - P_j(t)\}/\Delta t = b_{j-1}(t)P_{j-1}(t)$$
$$+ d_{j+1}(t)P_{j+1}(t) - [b_j(t) + d_j(t)]P_j(t) + o(\Delta t)/\Delta t.$$

(2.2.1) follows by letting $\Delta t \to 0$.

To compute $P_j(t)$ by using (2.2.1), one has to solve the equations (2.2.1). A general method for solving (2.2.1) is by way of PGF. The basic steps are given as follows:

i. Define the PGF $Q(z,t)$ of $X(t)$ given $X(t_0) = 1$ as

$$Q(z,t) = \sum_{j=0}^{\infty} z^j P_j(t), Q(z,t_0) = z.$$

ii. Use (2.2.1) to obtain an equation for $Q(z,t)$ and then solve the equation for $Q(z,t)$ under the initial condition $Q(z,t_0) = z$.

iii. Obtain $P_j(t)$ from $Q(z,t)$ by the formula:

$$(j!)P_j(t) = \{d^j Q(z,t)/dz^j\}_{z=0}.$$

To obtain $Q(z,t)$, one multiplies (2.2.1) on both sides by z^j and sums over j from 0 to ∞ to obtain:

$$\frac{\partial}{\partial t} Q(z,t) = \sum_{j=0}^{\infty} z^j \{(z-1)b_j(t) + (z^{-1}-1)d_j(t)\},$$

$$Q(z,t_0) = z. \tag{2.2.2}$$

The solution of (2.2.2) is only possible for some special cases. In the next section we shall consider three special cases which are relevant to processes of carcinogenesis.

2.3 SOME IMPORTANT SPECIAL STOCHASTIC BIRTH AND DEATH PROCESSESS FOR CARCINOGENESIS

Let $X(t), t \geq t_0$ be a general density-dependent birth-death process with birth rate $b_j(t)$ and death rate $d_j(t)$, as defined in the previous section. In this section we discuss some important special stochastic birth-death processes which are relevant to the processes of carcinogenesis.

2.3.1 The Poisson Processes

In this case, $b_j(t) = \lambda(t)$ and $d_j(t) = 0$ for all $j = 0, 1, 2, \ldots$. Hence, (2.2.1) reduces to:

$$\frac{d}{dt} P_j(t) = \lambda(t)[P_{j-1}(t) - P_j(t)], \ P_j(t_0) = \delta_{1j}, j = 0, 1, 2 \ldots \tag{2.3.1}$$

Equation (2.2.2) reduces to:

$$\frac{\partial}{\partial t} Q(z,t) = (z-1)\lambda(t)Q(z,t), Q(z,t_0) = z. \tag{2.3.2}$$

Or $\frac{\partial}{\partial t} \log Q(z,t) = (z-1)\lambda(t)$. Hence, $Q(z,t) = z \exp[(z-1)\Lambda(t_0,t)]$

$$= z \exp[-\Lambda(t_0,t)] \sum_{j=0}^{\infty} [z\Lambda(t_0,t)]^j/(j!),$$

where $\Lambda(t_0,t) = \int_{t_0}^{t} \lambda(x)dx$.

It follows that $P_0(t) = 0$ and for $j \geq 1$,

$$P_j(t) = \exp[-\Lambda(t_0,t)][\Lambda(t_0,t)]^{j-1}/(j-1)!.$$

Let $\kappa_j(t)$ be the jth cumulant of $X(t)$. Then $\kappa_1(t) = 1 + \Lambda(t_0,t)$ and for $j \geq 1, \kappa_j(t) = \Lambda(t_0,t)$.

2.3.2 The Feller-Arley Birth-Death Processes

In this case, $b_j(t) = jb(t)$ and $d_j(t) = jd(t)$. Hence (2.2.1) reduces to:

$$\frac{d}{dt} P_j(t) = -j[b(t) + d(t)]P_j(t) + (j-1)b(t)P_{j-1}(t)$$
$$+(j+1)d(t)P_{j+1}(t), P_j(t_0) = \delta_{1j}. \tag{2.3.3}$$

Since

$$\sum_{j=0}^{\infty} jz^j P_j(t) = z \frac{d}{dz} \sum_{j=0}^{\infty} z^j P_j(t) = z \frac{\partial}{\partial z} Q(z,t),$$

(2.2.2) reduces to:

$$\frac{\partial}{\partial t} Q(z,t) = (z-1)[zb(t) - d(t)] \frac{\partial}{\partial z} Q(z,t), Q(z,t_0) = z. \tag{2.3.4}$$

In Section (2.5), (2.3.4) is solved to yield the solution as:

$$Q(z,t) = 1 - (z-1)\{(z-1)\psi(t_0,t) - \phi(t_0,t)\}^{-1}, \tag{2.3.5}$$

where

$$\phi(t_0,t) = \exp\left\{-\int_{t_0}^{t} [b(x) - d(x)]dx\right\}$$

and

$$\psi(t_0,t) = \int_{t_0}^{t} b(y)\phi(t_0,y)dy.$$

Write for simplicity $\psi = \psi(t_0,t)$ and $\phi = \phi(t_0,t)$. Then, expanding $Q(z,t) = \sum_{j=1}^{\infty} z^j P_j(t)$, one obtains:

$$P_0(t) = 1 - [\phi + \psi]^{-1} \quad \text{and for} \quad j \geqslant 1,$$

$$P_j(t) = [\psi/(\phi + \psi)]^j \{\phi/[\psi(\phi + \psi)]\}.$$

(See Reference 17).

Let $C(z,t)$ be the cumulant generating function of $X(t)$. Then, using (2.3.5),

$$
\begin{aligned}
C(z,t) &= \log Q[\exp(z),t] \\
&= \log\{1 - [\exp(z) - 1][(e^z - 1)\psi(t_0,t) - \phi(t_0,t)]^{-1}\}, \quad (2.3.6)
\end{aligned}
$$

From (2.3.6), one may readily obtain the cumulants of $X(t)$ given $X(t_0) = 1$ by taking derivatives with respect to z and then setting $z = 0$. In particular, one obtains the first four cumulants as,

$$
\begin{aligned}
\kappa_1(t) &= [\phi(t_0,t)]^{-1} = \exp\left\{\int_{t_0}^t [b(x) - d(x)]dx\right\}, \\
\kappa_2(t) &= \{\phi(t_0,t) + 2\psi(t_0,t) - 1\}/\phi^2(t_0,t), \\
\kappa_3(t) &= \{\phi^2(t_0,t) + 3\phi(t_0,t)[2\psi(t_0,t) - 1] \\
&\quad + 2[3\psi^2(t_0,t) - 3\psi(t_0,t) + 1]\}\,\phi^3(t_0,t), \quad (2.3.7) \\
\kappa_4(t) &= \{\phi^3(t_0,t) + \phi^2(t_0,t)[2\psi(t_0,t) - 1] \\
&\quad + 12\phi(t_0,t)[3\psi^2(t_0,t) - 3\psi(t_0,t) + 1] \\
&\quad + 6[4\psi^3(t_0,t) - 6\psi^2(t_0,t) + 4\psi(t_0,t) - 1]\}[\phi(t_0,t)]^{-4}.
\end{aligned}
$$

If $b(t) - d(t) = \epsilon \exp(-\delta t)$ for some $\epsilon > 0$ and $\delta > 0$, then $E[X(t)|X(t_0) = 1] = \kappa_1(t) = \exp\{(\epsilon/\delta)[\exp(-\delta t_0) - \exp(-\delta t)]\}$ so that $X(t)$ is a Gompertz birth-death process. (see 17)

2.3.3 The Stochastic Logistic Birth-Death Processes

In this case, $S = \{0,1,2,\ldots,M\}, b_j(t) = d_{j+1}(t) = 0$ for $j \geqslant M$ and for $0 \leq j \leq M$, $b_j(t) = jb(t)[1 - (j/M)]$ and $d_j(t) = jd(t)[1 - (j/M)]$.

The Kolmogorov forward equations are:

$$
\begin{aligned}
\frac{d}{dt}P_j(t) &= -j[1 - (j/M)][b(t) + d(t)]P_j(t) \\
&\quad + (j - 1)\{1 - [(j - 1)/M]\}b(t)P_{j-1}(t) \\
&\quad + (j + 1)\{1 - [(j + 1)/M]\}d(t)P_{j+1}(t), j = 0,1\ldots,M - 1; \\
\frac{d}{dt}P_M(t) &= (M - 1)\{1 - [(M - 1)/M]\}b(t)P_{M-1}(t), P_j(t_0) = \delta_{1j}.
\end{aligned}
$$

Since

$$\sum_{j=0}^{\infty} j z^j [1 - (j/M)] P_j(t)$$

$$= z\{(1 - \frac{1}{M})\sum_{j=0}^{\infty} j z^{j-1} P_j(t)$$

$$-(\frac{z}{M})\sum_{j=0}^{\infty} j(j-1)z^{j-2} P_j(t)\}$$

$$= z\{(1 - \frac{1}{M})\frac{\partial}{\partial z} Q(z,t) - (\frac{z}{M})\frac{\partial^2}{\partial z^2} Q(z,t)\},$$

(2.2.2) reduces to

$$\frac{\partial}{\partial t} Q(z,t) = (z-1)[zb(t) - d(t)]\{(1 - \frac{1}{M})\frac{\partial}{\partial z}$$
$$- \frac{z}{M}\frac{\partial^2}{\partial z^2}\}Q(z,t), \qquad (2.3.8)$$

with $Q(z,t_0) = z$.

If $M \to \infty$, (2.3.8) reduces to (2.3.4). For $0 < M < \infty$, however, equation (2.3.8) is very difficult to solve. Tan and Piantadosi (21) have applied (2.3.8) to obtain the moments of $X(t)$ and derive a diffusion approximation for $Y(t) = X(t)/M$. By applying (2.3.8), Tan and Piantadosi (21) have shown that the first four cumulants of $X(t)$ satisfy, respectively, the following differential equations:

$$\frac{d}{dt} \kappa_1(t) = \epsilon(t)\{\kappa_1(t) - (1/M)[(\kappa_1(t))^2 + \kappa_2(t)]\}, \qquad (2.3.9)$$

where $\kappa_1(t_0) = 1$, and $\epsilon(t) = b(t) - d(t)$;

$$\frac{d}{dt} \kappa_2(t) = \kappa_2(t)\{2\epsilon(t)[1 - (2\kappa_1(t)/M)] - [\omega(t)/M]\}$$
$$+ \kappa_1(t)\omega(t)\{1 - [\kappa_1(t)/M]\} - 2\epsilon(t)[\kappa_3(t)/M], \qquad (2.3.10)$$

where $\kappa_2(t_0) = 0$ and $\omega(t) = b(t) + d(t)$;

$$\frac{d}{dt} \kappa_3(t) = 3\kappa_3(t)h_1(t) + \epsilon(t)h_2(t)$$
$$+ 3\omega(t)\kappa_2(t)\{1 - [2\kappa_1(t)/M]\} - [3\epsilon(t)\kappa_4(t)/M], \qquad (2.3.11)$$

where $\kappa_3(t_0) = 0, h_1(t) = \epsilon(t)\{1 - [2\kappa_1(t)/M]\} - [\omega(t)/M]$, and $h_2(t) = \kappa_1(t) - [\kappa_2(t) + \kappa_1^2(t) + 6\kappa_2^2(t)](1/M)$; and

$$\frac{d}{dt}\kappa_4(t) = \kappa_4(t)\{4h_1(t)$$
$$- [2\omega(t)/M]\} + h_3(t) - [4\epsilon(t)\kappa_5(t)/M], \quad (2.3.12)$$

where $\kappa_4(t_0) = 0$, and

$$h_3(t) = \omega(t)\{h_2(t) + 6\kappa_3(t) - 6[\kappa_2^2(t) + 2\kappa_1(t)\kappa_3(t)](1/M)\}$$
$$-4\epsilon(t)\{\kappa_2(t) - [6\kappa_3(t)\kappa_2(t) + \kappa_3(t) + 2\kappa_2(t)\kappa_1(t)](1/M)\}.$$

If $\kappa_i(t) = 0(M)$ for $i = 1, 2$, then

$$\frac{d}{dt}\kappa_1(t) = \epsilon(t)\{\kappa_1(t) - (1/M)[\kappa_1^2(t) + \kappa_2(t)]\}$$
$$\cong \epsilon(t)\kappa_1(t)[1 - (1/M)\kappa_1(t)], \kappa_1(t_0) = 1.$$

It follows that if $\kappa_i(t) = 0(M)$ for $i = 1, 2$, a close approximation to $\kappa_1(t)$ is given by:

$$\kappa_1(t) \cong \exp\left\{\int_{t_0}^t \epsilon(x)dx\right\} \cdot \{1 - (1/M)$$
$$+ (1/M)\exp\left[\int_{t_0}^t \epsilon(x)dx\right]\}^{-1}, \quad (2.3.13)$$

Equation (2.3.13) is the non-homogeneous logistic growth function. It is shown in Tan and Piantadosi (21) that when $b_j(t) = b_j$ and $d_j(t) = d_j$, (2.3.13) provides a very close approximation for most of the situations which correspond to the doubling time of bacteria and cell populations. Similarly, if $\kappa_i(t) = 0(M), i = 1, 2$ and if $\epsilon(t)\kappa_3(t) = 0(M)$,

$$\frac{d}{dt}\kappa_2(t) \cong \kappa_2(t)\{2h_1(t) + [\omega(t)/M]\} + \kappa_1(t)\omega(t)\{1 - [\kappa_1(t)/M]\}$$
$$= \kappa_2(t)f_1(t) + f_2(t), \quad (2.3.14)$$

where

$$\kappa_2(t_0) = 0, f_1(t) = 2h_1(t) + [\omega(t)/M]$$

and

$$f_2(t) = \kappa_1(t)\omega(t)\{1 - [\kappa_1(t)/M]\}.$$

From (2.3.14),

$$\kappa_2(t) \cong \exp\{\int_{t_0}^{t} f_1(x)dx\} \int_{t_0}^{t} f_2(x)\exp[-\int_{t_0}^{x} f_1(s)ds]dx \qquad (2.3.15)$$

$$= (1 - \frac{1}{M})g_1(t)\int_{t_0}^{t} \omega(x)g_2(x)dx,$$

where

$$g_1(t) = \exp\{\int_{t_0}^{t} [2\epsilon(x) - \frac{1}{M}\omega(x)]dx\}$$

$$\cdot \{1 - (1/M) + (1/M)\exp[\int_{t_0}^{t} \epsilon(x)dx]\}^{-4}$$

and

$$g_2(t) = \exp\{-\int_{t_0}^{t} [\epsilon(x) - \frac{1}{M}\omega(x)]dx\} \cdot \{1 - (1/M)$$

$$+ (1/M)\exp[\int_{t_0}^{t} \epsilon(x)dx]\}^2.$$

It is shown in Tan and Piantadosi (21) that when $b_j(t) = b$ and $d_j(t) = d$, (2.3.15) provides a very close approximation for many plausible values of the doubling time of bacteria and cell populations.

EXAMPLE (2.14) THE ONE-STAGE MODEL OF CARCINOGEN-ESIS In Chapter One and Example (2.13), it has been shown that under some infrequent situations, one stage is sufficient for the conversion of normal stem cells into malignant tumors.

Let $X(t)$ and $Y(t)$ be the numbers of normal stem cells and tumors at time t, respectively. I now proceed to show that under some general conditions, $Y(t)$ is in fact a nonhomogeneous Poisson process. Given below are these conditions:

i. $X(t)$ is very large for all $t \geq t_0$, where t_0 may be taken as the time of birth. Then one may take $X(t)$ as a deterministic function of t. This assumption is usually true for adult tissues; see reference 9.

ii. The tumor cells grow into tumors by following a nonhomogeneous Feller-Arley birth-death process with birth rate $b(t)$ and death rate $d(t)$. (This process is a stochastic Gompertz growth process if $\epsilon(t) = b(t) - d(t) = \epsilon \exp(-\delta t)$ for $\epsilon > 0, \delta > 0$.) Let M_D be the detectable size and $F_D(s,t)$ the probability that a tumor cell at time s is detectable as a tumor at or before time t given one tumor cell at time s. Denote by $P_D(s,t) = dF_D(s,t)/dt$; then $P_D(s,t) = P_{1M_D}(s,t)$ is given in Section (2.3.2) with $j = M_D$ and $t_0 = s$.

iii. During $[t, t + \Delta t]$, the probability that a normal stem cell at time t yields one normal stem cell and one tumor cell at time $t + \Delta t$ is $\alpha(t)\Delta t + o(\Delta t)$, where

$$\lim_{\Delta t \to 0} o(\Delta t)/\Delta t$$

iv. As in references (9) and (19), it is assumed that with probability one tumor cell will eventually develop into malignant tumors. From this it also follows that with probability one cancer tumor will not regress.

v. Each cell in the population goes through the processes independently of other cells.

Now, it is well known among geneticists that mutation rates are usually very small ($\alpha(t) \approx 10^{-6}$–$10^{-8}$). Since $X(t)$ is very large for all $t \geq t_0$ by assumption (i), one would expect that under (i) $X(t)\alpha(t) = \overline{\alpha}(t)$ is finite for all $t \geq t_0$. Under these conditions, it is at least approximately true from the theory of binomial distributions that during $[s, s + \Delta t]$, the number of new tumor cells arising from normal stem cells follows a Poisson distribution with parameter $\overline{\alpha}(t)\Delta t + o(\Delta t)$.

THEOREM (2.3.1)   Assume that $X(t)\alpha(t)$ is finite for all $t \geq t_0$ and that during $[s, s + \Delta t]$ the mutation process from normal stem cells to tumor cells follows a Poisson distribution with parameter $X(t)\alpha(t)\Delta t + o(\Delta t)$ independently, then under conditions (i)–(v) the PGF $Q(z,t)$ of $Y(t)$ given $Y(t_0) = 0$ is given by:

$$Q(z,t) = \exp\left\{(z - 1)\int_{t_0}^{t} X(u)\alpha(u)F_D(u,t)du\right\}.$$

Proof. To prove Theorem (2.3.1), partition the time interval $[t_0, t]$ by $I_j = [t_{j-1}, t_j), j = 1, \ldots, n - 1$ and $I_n = [t_{n-1}, t_n]$, where $t_j = t_0 + j\Delta t$ and $n\Delta t = t - t_0$ with $t = t_n$. Let $\psi(z; t, \Delta t)$ be the PGF of $Y(t)$ under the partition so that

$$Q(z,t) = \lim_{\Delta t \to 0} \psi(z; t, \Delta t).$$

Let M_j be the number of tumor cells arising from normal stem cells during the time interval $I_j, j = 1, \ldots, n$. Then, the conditional PGF of

$Y(t)$ given the partition and given $[M_j, j = 1, \ldots, n]$ is,

$$\prod_{j=1}^{n} [1 - F_D(t_{j-1}, t) + z F_D(t_{j-1}, t)]^{M_j} = \prod_{j=1}^{n} [(z - 1) F_D(t_{j-1}, t) + 1]^{M_j}.$$

Now, M_j follows a nonhomogeneous Poisson distribution with parameter $X(t_{j-1})\alpha(t_{j-1})\Delta t + o(\Delta t)$ independently. Taking expectation over $[M_j, j = 1, 2, \ldots, n]$, the conditional PGF of $Y(t)$ given the partition is

$$\psi(z; t, \Delta t) = \prod_{j=1}^{n} \exp \{-X(t_{j-1})\alpha(t_{j-1})\Delta t$$
$$+ X(t_{j-1})\alpha(t_{j-1})\Delta t[1 + (z - 1)F_D(t_{j-1}, t)] + o(\Delta t)\}.$$
$$= \exp \{(z - 1) \sum_{j=1}^{n} X(t_{j-1})\alpha(t_{j-1})F_D(t_{j-1}, t)\Delta t\}$$
$$+ (t - t_0)o(\Delta t)/\Delta t.$$

Letting $\Delta t \to 0$, one obtains:

$$Q(x, t) = \lim_{\Delta t \to 0} \psi(z; t, \Delta t)$$
$$= \exp \{(z - 1) \lim_{\Delta t \to 0} \sum_{j=1}^{n} X(t_{j-1})\alpha(t_{j-1})F_D(t_{j-1}, t)\Delta t\}$$
$$= \exp \{(z - 1) \int_{t_0}^{t} X(u)\alpha(u)F_D(u, t)du\}$$

Theorem (2.3.1) shows that if for all $t \geq t_0, X(t)$ is very large but the mutation rate $\alpha(t)$ is very small so that $\overline{\alpha}(t) = X(t)\alpha(t)$ is finite for all $t \geq t_0$, then one would expect that $Y(t)$ is a Poisson process. Note that if $Y(t)$ is Poisson with parameter

$$\Lambda(t_0, t) = \int_{t_0}^{t} X(u)\alpha(u)F_D(u, t)dt,$$

then

$$P_j(t) = \exp [-\Lambda(t_0, t)][\Lambda(t_0, t)]^j/(j!)$$

and the j th cumulant is $\kappa_j(t) = \Lambda(t_0, t), j = 1, 2, \ldots$. Further, the incidence function $\Lambda(t)$ is

$$\lambda(t) = -Q'(0, t)/Q(0, t)$$

$$= \{X(t)\alpha(t)F_D(t,t) + \int_{t_0}^t X(u)\alpha(u)[dF_D(u,t)/dt]du\}$$

$$= \int_{t_0}^t X(u)\alpha(u)P_D(u,t)du.$$

2.4 THE FILTERED POISSON PROCESSES

Many biomedical processes including mutagenesis and carcinogenesis are closely related to filtered Poisson processes. To model processes of carcinogenesis, it is essential to have a basic understanding of these processes. In this section I shall therefore proceed to give a brief account of these processes and illustrate how carcinogenesis models are related to these processes.

DEFINITION (2.4.1)   Let $Z(t), t \geqslant t_0$ be a stochastic process with state space $S = \{0, 1, 2, \ldots\}$. Then, $Z(t)$ is called a filtered Poisson process if the PGF $Q(z,t)$ of $Z(t)$ given $Z(t_0) = m_0$ for some integer $m_0 \geqslant 0$ is given by:

$$Q(z,t) = \exp\{\int_{t_0}^t X(u)[\phi(z;u,t) - 1]du\}, \tag{2.4.1}$$

where $X(u)$ is a nonnegative function and $\phi(z;u,t)$ is a PGF with z as dummy variable.

Definition (2.4.1) was proposed by Tan (16) in a multivariate setup. From (2.4.1), $Q(z,t)$ is known if $\phi(z;u,t)$ is known. Then one may use $Q(z,t)$ to find the probability distribution and the cumulants of $Z(t)$.

Some Properties of Filtered Poisson Processes

Let $Z(t), t \geqslant t_0$ be a filtered Poisson process with state space $S = \{0, 1, 2, \ldots\}$. Assume that the PGF of $Z(t)$ given $Z(t_0) = m_0(m_0 \geqslant 0$ an integer) is

$$Q(z,t) = \exp\{\int_{t_0}^t X(u)[\phi(z;u,t) - 1]du\},$$

where $\phi(z;u,t)$ is a PGF and $X(u)$ a nonnegative function.

Then the cumulant generating function $C(z,t)$ of $Z(t)$ given $Z(t_0) = m_0$ is $C(z,t) = \log Q[\exp(z),t]$, and the j th cumulant $\kappa_j(t)$ of $Z(t)$ given $Z(t_0) = m_0$ is,

$$\kappa_j(t) = \{d^j C(z,t)/dz^j\}_{z=0}.$$

In particular, the mean $\kappa_1(t)$ and the variance $\kappa_2(t)$ of $Z(t)$ given $Z(t_0) = m_0$ are given respectively by:

$$\kappa_1(t) = \int_{t_0}^{t} X(u)\omega_1(u,t)du \qquad (2.4.2)$$

and

$$\kappa_2(t) = \int_{t_0}^{t} X(u)[\omega_2(u,t) + \omega_1(u,t)]du, \qquad (2.4.3)$$

where $\omega_j(u,t) = \{d^j \phi(z;u,t)/dz^j\}_{z=1}.$

Let $P_j(t) = Pr\{Z(t) = j | Z(t_0) = m_0\}$ and let $V_j(u,t) = \{d^j \phi(z;u,t)/dz^j\}_{z=0}/(j!)$. Then $Q(z,t)$ can be written as

$$Q(z,t) = \exp\{-q_0(t) + \sum_{j=1}^{\infty} z^j q_j(t)\},$$

where

$$q_j(t) = (-1)^{\delta_{0j}} \int_{t_0}^{t} X(u)[V_j(u,t) - \delta_{0j}]du.$$

By Theorem (2.1.2), $P_j(t)$ is then given by the following iterative formula:

$$P_0(t) = \exp[-q_0(t)] \quad \text{and for} \quad j \geqslant 1,$$

$$P_j(t) = q_j(t)P_0(t) + \sum_{u=1}^{j-1} q_{j-u}(t)P_u(t)[(j-u)/j].$$

Using this iterative formula, Tan (14, 15) has computed the probabilities of the number of mutants in a cell population subjected to mutation and stochastic birth-death processes. Similarly, Tan and Gastardo (18) have computed the probabilities of the number of tumors in a two-stage model of carcinogenesis which is in essence a filtered Poisson process (see Chapter Three).

Some Examples

To illustrate the relationship between the Poisson processes and the filtered Poisson processes and to show the applications of filtered Poisson processes, given below are some specific examples from some biomedical problems.

EXAMPLE (2.15) Let $Z(t), t \geq 0$ be a stochastic process with state space $S = \{0, 1, 2, \ldots\}$. Suppose that $Z(t)$ is represented by:

$$Z(t) = \lim_{\Delta t \to 0} \sum_{j=0}^{n-1} \sum_{u=1}^{N(j\Delta t)} X_u[(n-j)\Delta t],$$

$$(2.4.4)$$

where $\quad Pr[X_u(0) = 1] = 1 \quad$ for all $\quad u = 1, 2, \ldots$.

In the representation (2.4.4), assume that the $N(j\Delta t)'s$ are Poisson variates with parameters $\beta(j\Delta t)\Delta t$ independently over different j values and that the $X_u[(n-j)\Delta t]'s$ are homogeneous Feller-Arley birth-death processes with birth rate b and death rate d independently over different u and j values. Then $Z(t)$ in (2.4.4) is a filtered Poisson process with PGF given by:

$$Q(z, t) = \exp \left\{ \int_0^t \beta(x)[f(z, t - x) - 1]dx \right\}, \quad \text{with} \quad t = n\Delta t,$$

where $f(z, t)$ is the PGF of a homogeneous Feller-Arley birth-death process $X(t), t \geq 0$ with birth rate b and death rate d given $X(0) = 1$; in fact $f(z, t) = 1 - \epsilon(z - 1)\{b(z - 1) - (zb - d) \times \exp(-\epsilon t)\}^{-1}$, where $\epsilon = b - d$. (See Section (2.3).)

To show that $Z(t)$ is a filtered Poisson process, note first that the PGF of $X_u[(n-j)\Delta t]$ given $X_u(0) = 1$ is $f[z; (n-j)\Delta t]$. Hence, given $N(j\Delta t) = m_j$, the PGF of

$$\sum_{u=1}^{N(j\Delta t)} X_u[(n-j)\Delta t] \quad \text{is} \quad [f(z; (n-j)\Delta t)]^{m_j}.$$

Since $N(j\Delta t)$ is Poisson with parameter $\beta(j\Delta t)\Delta t$, the unconditional PGF of

$$\sum_{u=1}^{N(j\Delta t)} X_u[(n-j)\Delta t] \quad \text{is} \quad \psi(z; t, \Delta t) = E\{f[z; (n-j)\Delta t]\}^{N(j\Delta t)}$$

$$= \exp\{-\beta(j\Delta t)\Delta t + \beta(j\Delta t)\Delta t f[z; (n-j)\Delta t]\}$$

$$= \exp\{\beta(j\Delta t)[f(z; (n-j)\Delta t) - 1]\Delta t\}.$$

Hence, the PGF of

$$\lim_{\Delta t \to 0} \sum_{j=0}^{n-1} \sum_{u=1}^{N(j\Delta t)} X_u[(n-j)\Delta t]$$

is

$$Q(z;t) = \lim_{\Delta t \to 0} \psi(z;t,\Delta t)$$

$$= \exp\{ \lim_{\Delta t \to 0} \sum_{j=0}^{n-1} \beta(j\Delta t)[f(z;(n-j)\Delta t) - 1]\Delta t \}$$

$$= \exp \{ \int_0^t \beta(u)[f(z;t-u) - 1]du \}.$$

To relate $Z(t)$ in (2.4.4) to some real practical problems, consider a very large population of bacteria consisting of mutants and nonmutants. Let $Y(t)$ and $Z(t)$ denote, respectively, the numbers of nonmutants and mutants at time t. Assume that for all $t \geqslant 0, Y(t)$ is very large and the mutation rate $\alpha(t)$ from nonmutant to mutant is very small so that $Y(t)\alpha(t) = \beta(t)$ is finite for all $t \geqslant 0$. Then one may assume that the number $N(j\Delta t)$ of new mutants arising from mutations of nonmutants during the time interval $I_j = [j\Delta t, (j + 1)\Delta t)$ is Poisson with parameter $\beta(j\Delta t)\Delta t$. Let $t = n\Delta t$ and partition the time interval $[0, t]$ into $I_j = [j\Delta t, (j + 1)\Delta t), j = 0, 1, \ldots, n - 1$. Then $X_u[(n-j)\Delta t]$ is the number of mutants at time t arising from the uth mutation of nonmutants during I_j by the homogeneous Feller-Arley birth-death process;

$$\sum_{u=1}^{N(j\Delta t)} X_u[(n-j)\Delta t]$$

is the total number of mutants at time t arising through the birth-death process from all mutations of nonmutants during I_j and

$$\sum_{j=0}^{n-1} \sum_{u=1}^{N(j\Delta t)} X_u[(n-j)\Delta t]$$

is the total number of mutants at time t arising through the birth-death process from mutations of nonmutants during $I_j, j = 0, 1, \ldots, n - 1$. Thus, $Z(t)$ is the total number of mutants arising through the birth-death process from mutations of nonmutants during $[0, t]$.

EXAMPLE (2.16) DRUG RESISTANT MODELS OF CANCER TUMORS In treating cancer tumors by anti-cancer drugs, a major difficulty is the development of resistance of tumor cells to anti-cancer drugs. Coldman and Goldie (1) and Ling et al (6) showed that most drug resistance of tumor cells was due to the development of resistant mutant cells. As illustrated by Tan and Brown (20), the problem of drug resistance of tumor cells can in fact be modeled by a filtered Poisson process. To illustrate, we assume that only one anti-cancer drug is applied. Then, in the population of cancer tumor cells, there are two types of tumor cells: The sensitive tumor cells and the resistant tumor cells.

Let $X(t)$ and $Y(t)$ be the numbers of sensitive tumor cells and resistant tumor cells at time t, respectively. Further, we make the following assumptions:

i.   At the starting time t_0, assume $Y(t_0) = 0$ and $X(t_0) = N_0$. If the tumor is detectable at t_0, then N_0 is very large, usually $N_0 \geq 10^7$.

ii. The sensitive tumor cells and the resistant tumor cells follow nonhomogeneous Feller-Arley birth-death processes for their growth with birth and death rates $(b_1(t), d_1(t))$ and $(b_2(t), d_2(t))$ respectively. If $b_i(t) - d_i(t) = \epsilon_i \exp(-\delta_i t)$ for some constants $\epsilon_i > 0, \delta_i > 0$, the processes become stochastic Gompertz birth-death processes (see 17).

iii. During $[t, t + \Delta t]$, the probability that a sensitive tumor cell at time t yields one sensitive cell and one resistant cell at time $t + \Delta t$ is $\alpha(t)\Delta t + o(\Delta t)$. It is assumed that there is no backward mutation from resistant tumor cells to sensitive tumor cells.

iv. All cells go through the processes independently of all other cells.

In most of the practical situations (see 1, 6, 7, 9), $\alpha(t)$ is very small but $N_0\alpha(t)$ is finite for all $t \geq t_0$; further $b_i(t) \geq d_i(t) \geq 0$ and $b_1(t) - d_1(t) - \alpha(t) \geq 0$. It follows that $X(t)\alpha(t)$ is finite for all $t \geq t_0$ and by the theory of binomial distribution, one has that to the order $0(N_0^{-1})$, the mutation process from sensitive tumor cells to resistant tumor cells during $[t, t + \Delta t]$ follows a non-homogeneous Poisson process with parameter $X(t)\alpha(t)\Delta t + o(\Delta t)$. Thus, one may further assume a non- homogeneous Poisson process for the mutation from sensitive tumor cells to resistant tumor cells (see reference 20).

THEOREM (2.4.1) Assume that $X(t)\alpha(t) = \beta(t)$ is finite for all $t \geq t_0$ and that during $[t, t + \Delta t]$, the mutation process from sensitive tumor cells to resistant tumor cells follows a non-homogeneous Poisson process with parameter $\beta(t)\Delta t + o(\Delta t)$, independently. Then the PGF of $Y(t)$ given $(Y(t_0) = 0, X(t_0) = N_0)$ is

$$Q(z,t) = \exp\{\int_{t_0}^{t} \beta(x)[f(z;x,t) - 1]dx\},$$

where

$$f(z;s,t) = 1 - (z-1)\{(z-1)\psi(s,t) - \phi(x,t)\}^{-1}$$

with

$$\phi(s,t) = \exp\{-\int_{s}^{t} [b_2(x) - d_2(x)]dx\} \quad \text{and}$$

$$\psi(s,t) = \int_{s}^{t} b_2(y)\phi(s,y)dy.$$

Proof. To prove Theorem (2.4.1), partition the time interval $[t_0,t]$ by $[t_0,t] = \cup_{j=1}^{n}L_j$, where $L_j = [t_{j-1},t_j), j = 1,\ldots,n-1$ and $L_n = [t_{n-1},t_n]$ with $t = t_n, t_j = t_0 + j\Delta t, j = 1,\ldots,n$. Let $\psi(z;t,\Delta t)$ be the PGF of $Y(t)$ given the partition so that

$$Q(z,t) = \lim_{\Delta t \to 0} \psi(z;t,\Delta t).$$

Let M_j be the number of resistant tumor cells arising by mutation from sensitive tumor cells during $L_j, j = 1,\ldots,n$. Since $f(z;s,t)$ is the PGF of resistant tumor cells at time t given a resistant mutant which arises at time s from sensitive tumor cells, the conditional PGF of $Y(t)$ given the partition $[L_j, j = 1,\ldots,n]$ and given $[M_j, j = 1,\ldots,n]$ is $\prod_{j=1}^{n}[f(z;s_j,t)]^{M_j}$, where s_j with $t_{j-1} \le s_j < t_j$ represents the time for the arrival of a resistant mutant from sensitive tumor cells during L_j. Since M_j follows a non-homogeneous Poisson process with parameter $X(t)\alpha(t)\Delta t + o(\Delta t) = \beta(t)\Delta t + o(\Delta t)$ independently for $j = 1,\ldots,n$, by taking expectation over $[M_j, j = 1,\ldots,n]$, one obtains:

$$\psi(z;t,\Delta t) = \prod_{j=1}^{n} \exp\{-X(t_{j-1})\alpha(t_{j-1})\Delta t$$

$$+ f(z;s_j,t)X(t_{j-1})\alpha(t_{j-1})\Delta t + o(\Delta t)\}$$

$$= \exp\{\sum_{j=1}^{n}[f(z;s_j,t) - 1]X(t_{j-1})\alpha(t_{j-1})\Delta t + (t-t_0)o(\Delta t)/\Delta t\}.$$

Letting $\Delta t \to 0$ and noting that $t - t_0 = n\Delta t$, one obtains:

$$Q(z,t) = \lim_{\Delta t \to 0} \psi(z;t,\Delta t) = \exp\{\int_{t_0}^{t} X(s)\alpha(s)[f(z;s,t) - 1]ds\}.$$

Theorem (2.4.1) shows that under conditions (i)–(vi), $Y(t)$ is in fact a filtered Poisson process. (See reference 20).

EXAMPLE (2.17)   THE MULTIEVENT MODEL AS A FILTERED POISSON PROCESS   As shown in Chapters 3 and 6, the two stage model as proposed by Moolgavkar and Venzon (8) and the multievent model as proposed by Tan, Chu and Brown (23) are in fact filtered Poisson processes under some general conditions. For details, see Chapters 3 and 6.

2.5   SOLUTIONS OF SOME FIRST ORDER PARTIAL DIFFERENTIAL EQUATIONS AND RICATTI EQUATIONS

In developing stochastic models of carcinogenesis, one would very often encounter first order partial differential equations as given in (2.5.1) and the Ricatti equations as given in (2.5.7). In this section we thus illustrate how to solve these types of equations.

2.5.1   The Solution of Some First Order Partial Differential Equations

Consider the first order partial differential equations of the form:

$$\frac{\partial}{\partial t} \phi(z_1,\ldots,z_k,t) + \sum_{j=1}^{k} a_j(z_1,\ldots,z_k,t) \frac{\partial}{\partial z_j} \phi(z_1,\ldots,z_k,t) = 0 \quad (2.5.1)$$

with initial condition

$$\phi(z_1,\ldots,z_k,t_0) = a_0(z_1,\ldots,z_k).$$

As shown in most texts of partial differential equations (see for example reference 13, p. 696), the solution $\phi(z_1,\ldots,z_k,t)$ of equation (2.5.1) is obtained by the following procedures:

1. Consider the auxiliary equations

$$\frac{d}{dt} z_j = a_j(z, \ldots, z_k, t), j = 1, \ldots, k. \tag{2.5.2}$$

Solve (2.5.2) to obtain

$$g_j(z_1, \ldots, z_k, t) = C_j,$$

where the $C_j's$ are constants independent of t

2. The general solution $\phi(z_1, \ldots, z_k, t)$ is given by:

$$\phi(z_1, \ldots, z_k, t) = \zeta[g_1(z_1, \ldots, z_k, t), \ldots g_k(z_1, \ldots, z_k, t)]$$

where $\zeta(y_1, \ldots, y_k)$ is an arbitrary function of $(y_1, \ldots, y_k)$

3. Determine the solution by the initial condition

$$\phi(z_1, \ldots, z_k, t_0) = \zeta[g_1(z_1, \ldots, z_k, t_0), \ldots, g_k(z_1, \ldots, z_k, t_0)]$$
$$= a_0(z_1, \ldots z_k).$$

EXAMPLE (2.18)   The first order partial differential equation is

$$\frac{\partial}{\partial t} \phi(z, t) = [(z - 1)][(z - 1)b(t) + \epsilon(t)] \frac{\partial}{\partial z} \phi(z, t), \phi(z, t_0) = z. \tag{2.5.3}$$

i. For (2.5.3), the auxiliary equations is

$$\frac{d}{dt} z = \frac{d}{dt} (z - 1) = -(z - 1)[(z - 1)b(t) + \epsilon(t)].$$

Dividing both sides by $(z - 1)^2$, we obtain:

$$\frac{d}{dt} (z - 1)^{-1} = b(t) + (z - 1)^{-1} \epsilon(t); \text{ or}$$

$$\frac{d}{dt} \psi(z) - \psi(z)\epsilon(t) = b(t), \quad \text{where} \quad \psi(z) = (z - 1)^{-1}.$$

Multiplying the above equation on both sides by $V(t_0, t) = exp[-\int_{t_0}^{t} \epsilon(x)dx]$, we obtain

$$\frac{d}{dt} \{\psi(z)V(t_0, t)\} = b(t)V(t_0, t).$$

It follows that

$$g(z,t) = (z-1)^{-1}exp\left[-\int_{t_0}^t \epsilon(x)dx\right]$$

$$-\int_{t_o}^t b(y)exp\left[-\int_{t_0}^y \epsilon(x)dx\right]dy = C,$$

where C is a constant independent of t.

ii. The general solution is

$$\phi(z,t) = \zeta[g(z,t)],$$

where $\zeta(\ )$ is arbitrary

iii. Putting $t = t_0$, we have $g(z,t_0) = (z-1)^{-1}$ so that

$$\phi(z,t_0) = z = \zeta[(z-1)^{-1}].$$

It follows that $\zeta(x) = [1 + x^{-1}]$.
Hence

$$\phi(z,t) = 1 + \{g(z,t)\}^{-1},$$

$$= 1 - (z-1)\{U(t_0,t)(z-1) - V(t_0,t)\}^{-1}$$

where $V(t_0,t) = \exp\{-\int_{t_0}^t \epsilon(x)dx\}$ and $U(t_0,t) = \int_{t_0}^t b(y)V(t_0,y)dy$.

Example (2.19).

The partial differential equation is

$$\frac{\partial}{\partial t}\phi(y,z,t) = \{by^2 - [b + d + (1-z)\alpha]y + d\}\frac{\partial}{\partial y}\phi(y,z,t),$$

with $\phi(y,z,0) = y$ \hfill (2.5.4)

i. The auxiliary equations are

$$\frac{d}{dt}z = 0 \hspace{3cm} (2.5.5)$$

and

$$\frac{d}{dt}y = -\{by^2 - [b + d + (1-z)\alpha]y + d\}. \hspace{1cm} (2.5.6)$$

From (2.5.5) $z = C_1$, a constant independent of t.
Let $h(z) = [b+d+(1-z)\alpha]^2 - 4bd$ and let β_1 and β_2 be defined by

$$2b\beta_1 = [b + d + (1-z)\alpha] - [h(z)]^{\frac{1}{2}}$$

and

$$2b\beta_2 = [b + d + (1 - z)\alpha] + h(z)^{\frac{1}{2}}.$$

Then $h(z) > 0, \beta_2 > \beta_1$ and β_1 and β_2 are the two real roots of $by^2 - [b + d + (1 - z)\alpha]y + d = 0$.
Writing (2.5.6) as

$$\frac{d}{dt}y = \frac{d}{dt}(y - \beta_2) = -b(y - \beta_1)(y - \beta_2)$$

$$= -b(y - \beta_2)[y - \beta_2 + \beta_2 - \beta_1],$$

one may proceed as in Example (2.18) to obtain the solution of (2.5.6) as

$$g(y,z,t) = [(\beta_2 - \beta_1)(y - \beta_2)]^{-1}\{(\beta_2 - y) + (y - \beta_1)$$

$$\times \exp[-b(\beta_2 - \beta_1)t]\} = C_2,$$

where C_2 is a constant independent of t.

ii. The general solution of $\phi(y,z,t)$ is

$$\phi(y,z,t) = \xi[C_1, g(y,z,t)] = \zeta[g(y,z,t)],$$

where $\zeta(\ )$ is an arbitrary function.

iii. Putting $t = 0$, $g(y,z,0) = (y - \beta_2)^{-1}$. Hence, $\phi(y,z,0) = y = \zeta[(y - \beta_2)^{-1}]$ or $\zeta(\omega) = \beta_2 + \omega^{-1}$.

It follows that

$$\phi(y,z,t) = \zeta[g(y,z,t)] = \beta_2 + [g(y,z,t)]^{-1}$$

$$= \beta_2 + (\beta_2 - \beta_1)(y - \beta_2)\{(\beta_2 - y)$$

$$+ (y - \beta_1)\exp[-b(\beta_2 - \beta_1)t]\}^{-1}.$$

$$= \{\beta_1(\beta_2 - y) + \beta_2(y - \beta_1)\exp[-b(\beta_2 - \beta_1)t]\}$$

$$\times \{(\beta_2 - y) + (y - \beta_1)\exp[-b(\beta_2 - \beta_1)t]\}^{-1}.$$

EXAMPLE (2.20)   The partial differential equation is

$$\frac{\partial}{\partial t}\phi(t_0,t) = x_2[(x_2 - 1)b_2(t) + (x_3 - 1)\alpha_2(t)]\frac{\partial}{\partial x_2}\phi(t_0,t), \phi(t_0,t_0) = x_2.$$

The auxiliary equation is

$$\frac{d}{dt}x_2 = -x_2[x_2b_2(t) - b_2(t) + (x_3 - 1)\alpha_2(t)].$$

Or, equivalently, $\frac{d}{dt}x_2^{-1} = b_2(t) + [(x_3 - 1)\alpha_2(t) - b_2(t)]x_2^{-1};$ or

$$\frac{d}{dt}z + z[b_2(t) - (x_3 - 1)\alpha_2(t)] = b_2(t), \quad \text{where} \quad z = x_2^{-1};$$

or

$$\frac{d}{dt}\{z\exp[\int_{t_0}^t (b_2(y) - (x_3 - 1)\alpha_2(y))dy]\}$$

$$= b_2(t)\exp\{\int_{t_0}^t [b_2(y) - (x_3 - 1)\alpha_2(y)]dy\}.$$

This gives,

$$g(x_2, x_3; t_0, t) = x_2^{-1}\xi(x_3; t_0, t) - \eta(x_3; t_0, t) = \text{constant, where}$$

$$\xi(x_3; t_0, t) = \exp\{\int_{t_0}^t [b_2(x) + (1 - x_3)\alpha_2(x)]dx\} \quad \text{and}$$

$$\eta(x_3; t_0, t) = \int_{t_0}^t b_2(u)\xi(x_3; t_0, u)du.$$

Hence the general solution is $\phi(t_0, t) = \zeta[g(x_2, x_3; t_0, t)]$, where $\zeta(x)$ is arbitrary. Since $g(x_2, x_3; t_0, t_0) = x_2^{-1}$ and $\phi(t_0, t_0) = \zeta(x_2^{-1}) = x_2$, so, $\phi(t_0, t) = [g(x_2, x_3; t_0, t)]^{-1} = x_2\{\xi(x_3; t_0, t) - x_2\eta(x_3; t_0, t)\}^{-1}$.

2.5.2 Solution of a Ricatti Equation Arising From Cancer Research

In developing two-stage models of carcinogenesis, for the homogeneous cases one would encounter the following Ricatti equation with initial condition $\phi(0) = x_2$, where x_2 and x_3 are dummy variables:

$$\frac{d}{dt}\phi(t) = b_2\phi^2(t) + [\alpha_2 x_3 - (b_2 + d_2 + \alpha_2)]\phi(t) + d_2, \quad (2.5.7)$$

where $\alpha_2 > 0$ and $b_2 \geq d_2 \geq 0$.

To solve (2.5.7), let y_2 and y_1 with $y_2 > y_1$ be given by $2b_2y_2 = (b_2 + d_2 + \alpha_2 - \alpha_2 x_3) + h(x_3)$ and $2b_2y_1 = (b_2 + d_2 + \alpha_2 - \alpha_2 x_3) - h(x_3)$ with $h(x_3) = [(b_2 + d_2 + \alpha_2 - \alpha_2 x_3)^2 - 4b_2d_2]^{\frac{1}{2}}$. (Note $h(x_3) \geq 0$ so that $y_2 > y_1$ are real numbers for all x_3). Then, equation (2.5.7) becomes,

$$\frac{d}{dt}\phi(t) = b_2[\phi(t) - y_2 + y_2 - y_1][\phi(t) - y_2].$$

Dividing both sides by $[\phi(t) - y_2]^2$, then, with $\eta(t) = [\phi(t) - y_2]^{-1}$, one obtains:

$$\frac{d}{dt}\eta(t) + b_2(y_2 - y_1)\eta(t) = -b_2, \eta(0) = [\phi(0) - y_2]^{-1} = (x_2 - y_2)^{-1}.$$

Thus,

$$\frac{d}{dt}\{\eta(t)\exp[b_2(y_2 - y_1)t]\} = -b_2\exp[b_2(y_2 - y_1)t].$$

It follows that

$$\eta(t)\exp[b_2(y_2 - y_1)t] = -b_2\int_0^t \exp[b_2(y_2 - y_1)z]dz + (x_2 - y_2)^{-1}$$

$$= (y_2 - y_1)^{-1}\{1 - \exp[b_2(y_2 - y_1)t]\} + (x_2 - y_2)^{-1}$$

$$= [(y_2 - y_1)(x_2 - y_2)]^{-1}\{(x_2 - y_1) + (y_2 - x_2)\exp[b_2(y_2 - y_1)]t\}.$$

Since $\eta(t) = [\phi(t) - y_2]^{-1}$, so

$$\phi(t) = y_2 + (y_2 - y_1)(x_2 - y_2)\exp[b_2(y_2 - y_1)t]\{(x_2 - y_1)$$
$$+ (y_2 - x_2)\exp[b_2(y_2 - y_1)t]\}^{-1}$$
$$= \{y_2(x_2 - y_1) + y_1(y_2 - x_2)\exp[b_2(y_2 - y_1)t]\}\{(x_2 - y_1)$$
$$+ (y_2 - x_2)\exp[b_2(y_2 - y_1)t]\}^{-1}. \qquad (2.5.8)$$

2.6 CONCLUSIONS AND SUMMARY

For developing mathematical theories for carcinogenesis models, in this chapter we summarize some basic mathematical tools and some basic stochastic processes which will be used in later chapters. To start this chapter, we first define in Section (2.1) the incidence function and the probability generating function of the number of tumor cells and illustrate how to derive the incidence functions from the probability generating functions. Relevant to biomedical problems, a procedure for computing the probabilities from the probability generating functions is given in Section (2.2).

To describe the growth of stem cells, intermediate cells and tumor cells stochastically, in Section (2.2) we introduce the nonhomogeneous, density-dependent stochastic birth-death processes and illustrate how to derive the probability generating functions of these processes. Then in Section (2.3) we present three important stochastic birth-death processes-the nonhomogeneous Poisson process, the nonhomogeneous Feller-Arley birth-death process and the stochastic logistic birth-death process, and derive the probability generating functions and the cumulants for these processes. As an example, it is shown that under some general conditions, the one-stage model of carcinogenesis can be described by a nonhomogeneous Poisson process.

Since most carcinogenesis models are in fact filtered Poisson processes, in Section (2.4) we thus define the filtered Poisson processes and give some of its properties. It is shown that under some general conditions the stochastic drug resistant models in cancer treatment and the stochastic multievent models of carcinogenesis are in fact filtered Poisson processes.

REFERENCES

1. Coldman, A. J. and Goldie, J. H., A model for the resistance of tumor cells to cancer chemotherapeutic agents, Math. Biosciences **65** (1983), 291– 307.
2. Cox, D.R., Regression models and life tables (with discussion), Jour Roy. Statis. Society **B 74** (1972), 187-220.
3. Eisen, M., Mathematical Models In Cell Biology and Cancer Chemo-therapy, Lecture Notes in Biomathematics **No. 30** (1979), Springer-Verlag, New York.
4. Laird, A.K., Dynamics of tumor growth, Brit. J. Cancer **18** (1964), 490-502.
5. Laird, A.K., Dynamics of growth in tumors and in normal organisms, "Human Tumor Cell Kinetics Monograph NCI/NIH," Bethesda, MD, 1969, pp. 15-28.
6. Ling, V. et al, Quantitative genetic analysis of tumor progression, Cancer and Metastasis Review **4** (1985), 173-194.
7. Moolgavkar, S. H., Carcinogenesis modeling: From molecular biology to epidemiology, Ann. Rev. Public Health **7** (1986), 151–169.
8. Moolgavkar, S.H. and Venzon, D.J., Two event models for car-cinogenesis: Incidence curves for childhood and adult cancer, Math Biosciences **47** (1979), 55-77.
9. Moolgavkar, S.H, and Knudson, A.G., Mutation and cancer: A model for human carcinogenesis, Jour. Nat. Cancer Inst. **66** (1981), 1037-1052.
10. Moran, P., Random processes in genetics, Proc. Comb. Philos. Soc. **58** (1958), 299-311.

11. Parzen, E., "Stochastic Processes," Holden-Day, Inc., San Francisco, 1962.
12. Simpson-Herren, L. and Lloyd, H.H., Kinetic parameters and growth curves for experimental tumor systems, Cancer Chemother. Rep. Part 1 **54** (1970), 143-174.
13. Syski, R., "Introduction to Congestion Theory in Telephone Systems," Oliver and Boyd, Edinburgh, 1960.
14. Tan, W.Y., On the probability distribution of the numbers of mutants in cell populations, SIAM Jour. Appl. Math **42** (1982), 719-730.
15. Tan, W.Y., On probability distribution of the numbers of mutants at the hypoxanthine-quanine phosphoribosal transferase locus in Chinese hamster ovary cells, Math. Biosciences **67** (1983), 175-192.
16. Tan, W.Y., Multivariate filtered Poisson processes and applications to multivariate stochastic modeling of mutagenicity and carcinogenesis, Utilitas Mathematica **26** (1984), 63-78.
17. Tan, W.Y., A stochastic Gompertz birth-death process, Statist and Prob. Lett **4** (1986), 25-28.
18. Tan, W.Y. and Gastardo, M.T., On the assessment of effects of environmental agents on cancer tumor development by a two-stage model of carcinogenesis, Math Biosciences **74** (1985), 143-155.
19. Tan, W.Y. and Brown, C.C., A nonhomogeneous two-stage model of carcinogenesis, Math. Modelling **9** (1987), 631-642.
20. Tan, W.Y.and Brown, C.C., Cancer chemotherapy with immunostimulation: A nonhomogeneous stochastic model for drug resistance under immunization. I. One drug case, Math. Biosciences **97** (1989), 145–160.
21. Tan, W.Y. and Piantadosi, S., On stochastic growth processes with application to stochastic logistic growth, To appear in Statistica Sinica (1991).
22. Tan, W.Y. and Tou, C., Maximum likelihood estimation of genetic parameters in cell populations, Biom. J. **29** (1987), 297-218.
23. Tan, W. Y., Chu, K. and Brown, C. C., The stochastic multievent model for cancer: A stochastic multistage model with birth and death processes, National Cancer Institute/NIH, DCPC, Bethesda, MD, 1985.

3

Two-Stage Models of Carcinogenesis

In the past years, the two-stage models of carcinogenesis as proposed by Moolgavkar and Knudson (49) have received considerable attention among cancer epidemiologists and cancer researchers and have been suggested as a major model for assessing risks of environmental agents (50, 51, 58, 70, 73,74). This is mainly due to the fact that the model is complex enough to involve genetic changes and cell proliferation of normal stem cells and initiated cells; yet it is simple enough to be applicable to human incidence data (48,49), and to case control human epidemiological data (35). Because of its importance, in this chapter I proceed to develop in great detail the mathematical theories for this model and illustrate how it can be used to interpret animal experimental results and human epidemiological studies.

In Section (3.1), a brief survey of the literature and some biological evidence supporting the models are presented. The models together with some results of probability generating functions are given in Section (3.2). Some important special cases of the models are discussed in Section (3.3). Given the models, in Section (3.4) we proceed to develop age-specific incidence functions of tumors. Section (3.5) is devoted to the derivation of probability distributions of numbers of tumors and the means and variances of the numbers of tumors.

To demonstrate the usefulness of the models, in Section (3.6) we illustrate how to use the two-stage models to interpret some important biological phenomena in carcinogenesis. Finally in Section (3.7), we demonstrate how the two-stage models may be used to assess risks of environmental agents.

3.1 NOTES ON LITERATURE AND BIOLOGICAL EVIDENCE

The two-stage model of carcinogenesis that incorporates the kinetics of cell birth and death was first proposed by Armitage and Doll in 1957 by assuming deterministic exponential growth for normal cells and intermediate cells (2). Two stage models of carcinogenesis which assume stochastic growth for intermediate cells were then proposed by Kendall (28), Neyman and Scott (52) and most recently by Moolgavkar and Venzon (47) and Moolgavkar and Knudson (49) (see also 44, 45, 46). To broaden its applications, Tan and Gastardo (68), Tan and Brown (69) and Moolgavkar, Dewanji and Venzon (50) have extended this two-stage model into nonhomogeneous cases.

On the application side, Moolgavkar, Day and Stevens (48) applied the two-stage model to evaluate risk factors of breast cancer of women. Moolgavkar and Knudson (49) illustrated how their two-stage model could fit incidence functions of all human cancers by assuming different growth patterns for normal stem cells of different tissues. Tan and Gastardo (68) have illustrated how to use the above two-stage model to assess effects of environmental agents which can be classified as initiators and promoters. Krailo, Thomas and Pike (35) have illustrated how the above two-stage model can be fitted into case control data of breast cancers of women in the city of Los Angeles. Tan and Singh (70) have aplied the Moolgavkar-Knudson two stage model and the Tan-Gastardo extension to assess effects of metabolism of carcinogens on cancer tumor development. Recently, Thorslund, Brown and Charnley (74) and Moolgavkar, Dewanji and Venzon (50) have proposed using the above two-stage model to assess the risk of environmental agents. Given below are some biological evidence which supports the two-stage model as proposed by Moolgavkar and Knudson (49). (In what follows, we refer to the two-stage model proposed by Moolgavkar and Venzon (47) and Moolgavkar and Knudson (49) as the MVK model unless otherwise stated).

1. The MVK two-stage model appears to fit incidence functions of all human cancers while other models such as the Armitage -Doll multistage model (Armitage and Doll (3)) can only fit human incidence data of some cancers (49).

2. The discovery of antioncogenes (29,33,78) provides biological support for the MVK model. As noted by Moolgavkar (46), pedigree analysis has shown that human cancers in some families are transmitted in an autosomal dominant fashion. Cytogenetic analysis of these hereditary cancers have revealed that particular genes are deleted. Thus, in contrast to oncogenes, it is the inactivation of these antioncogenes that leads to malignancy. Examples of antioncogenes include the retinoblastoma Rb gene on chromosome 13q14(12,13,18,23,25) and the Wilm's tumor Wm gene on chromosome 11p13 (22,34,54).

3. Evidence to support the MVK two-stage model is also provided by the discovery of two complementing classes of oncogenes: One class is referred to as the"immortalization class" including cellular oncogenes myc, myb and viral oncogenes SV-40 large T, polyoma large T and adenovirus ElA, while the other class is referred to as the"transforming class" and includes viral oncogene, polyoma middle T and cellular oncogenes ras, src, erbB, neu, fms, fes/fps, yes, mil/raf, mos and abl (5,8,10,30,36,40,78.) It has been demonstrated that cotransfection of one oncogene from the immortalization class with an oncogene from the transforming class would lead to tumorigenic conversion of normal stem cells; but oncogenes from only one class would in general not induce conversion of normal stem cells (36).

4. The MVK two-stage model provides satisfactory explanations for many biological phenomena such as initiation and promotion (44,68), hormone effects in breast cancer (45) and many others (46). Many of these phenomena cannot be explained by the Armitage–Doll multistage model.

Working with the childhood eye cancer retinoblastoma, Knudson (32) appeared to be the first to classify the two-stage models into hereditary two-stage models and nonhereditary two-stage models. This was due to the fact that while the second event occurs only in somatic cells, the first event can occur either in germline cells or in somatic cells. In retinoblastoma, both hereditary cancers and nonhereditary cancers have been observed (12,13). This may help explain why cancer cases often cluster among relatives.

3.2 A GENERAL TWO-STAGE MODEL OF CARCINOGENESIS

The two-stage model of carcinogenesis as proposed by Moolgavkar and Knudson (49) assumes that a cancer tumor develops from a single normal stem cell by clonal expansion and views carcinogenesis as the end result of two discrete, heritable and irreversible events in normal stem cells. Each event occurs during a single cell division. It differs from the classical Armitage–Doll multistage model in that the normal stem cells and intermediate cells are subjected to cell proliferation and cell differentiation and death (see Figure 3.1). A distinct feature of this model is that the first event may occur either in germline cells or in somatic cells but the second event always occurs in somatic cells.

According to this model, there are three types of cancer cells: The normal stem cells, the intermediate cells (or initiated cells) and the tumor cells. Further, with probability one, a tumor cell will develop into a malignant tumor. We denote by $N(t), I(t)$ and $T(t)$ the numbers of normal stem cells (N cells), intermediate cells (I cells) and cancer tumors at time t, respectively. We shall proceed to find the age specific incidence

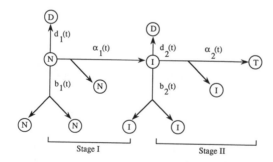

Figure 3.1 Two stage model of carcinogenesis.
N, I, T stand for normal stem cell, intermediate cell and tumor cell, respectively, D stands for death.

function of tumors and hence the probability distribution of time required for a normal stem cell to develop into a tumor.

3.2.1 Some Basic Assumptions and Consequences

Some Basic Assumptions

To solve many problems related to the above model, we make the following assumptions. Note that these assumptions are the same assumptions made in Tan and Brown (69) and are less restrictive than the assumptions made in Moolgavkar and Knudson (49) since we do not assume constant mutation rates and constant birth and death rates for normal stem cells and initiated cells.

    1. At the starting time t_0, there are N_0 normal stem cells. If an organ is well developed by time t_0, N_0 is usually very large ($N_0 \approx 10^6$–10^9). Hence if t_0 is taken as the time of birth, then one may assume large N_0 for most organs since at the time of birth, most organs are well developed.

    2. During $[t, t + \Delta t]$, the probability that a normal stem cell at time t yields one normal stem cell and one initiated cell at time $t + \Delta t$ is $\alpha_1(t)\Delta t + o(\Delta t)$, where

$$\lim_{\Delta t \to 0} o(\Delta t)/\Delta t = 0.$$

Similarly, the probability that an initiated cell at time t yields one initiated cell and one tumor cell at time $t + \Delta t$ is $\alpha_2(t)\Delta t + o(\Delta t)$.

    3. Normal stem cells follow a density dependent nonhomogeneous general birth and death process with birth rate $b_{1j}(t)$ and death rate $d_{1j}(t)(j = 0, 1, 2, \ldots)$ for their cell proliferation and differentiation. (See Section (2.2) of Chapter Two.) Note that for this general birth and death process it is feasible to adopt either Feller-Arley birth-death processes, or stochastic logistic birth-death processes (71) or stochastic Gompertz birth-death processes (67) for cell proliferation and differentiation of normal stem cells.

    4. The initiated cells are assumed to follow a nonhomogeneous Feller-Arley birth-death process with birth rate $b_2(t)$ and death rate $d_2(t)$ for cell proliferation and differentiation. Note that if $b_2(t) - d_2(t) = \epsilon_2 \exp(-\delta_2 t)$ for some constants $\epsilon_2 > 0$ and $\delta_2 > 0$, the nonhomogeneous birth-death Feller-Arley process then reduces to a stochastic Gompertz birth-death process (67).

5. As shown by Moolgavkar and Knudson (49), the time period for the development of a clinically detectable cancer tumor from a tumor cell is short relative to the time for the transformation process of normal stem cells into cancer tumors. As a close approximation, we thus assume that with probability one a tumor cell would develop into a cancer tumor. Further, one may ignore the random variation for the time elapsed between the birth of tumor cells and the development of cancer tumors.

6. The birth-death processes and the mutation processes are independent of one another and each cell goes through the above processes independently of other cells.

REMARKS

1. If $\alpha_i(t) = \alpha_i, i = 1, 2$ and $b_2(t) = b_2$ and $d_2(t) = d_2$ are independent of t, and if the normal stem cells follow some deterministic law of growth for their proliferation and differentiation, the above model reduces to the model considered by Moolgavkar and Knudson (49).

2. In most cases $\alpha_1(t)$ is very small (10^{-6}–10^{-8}). If N_0 is very large and if $b_{1j}(t) \geq d_{1j}(t)$, then $N(t)$ is expected to be very large for all $t \geq t_0$. In these cases the effects of random variations on cell proliferation and differentiation of normal stem cells are expected to be very small, so that one may assume some deterministic law of growth for cell proliferation and differentiation of normal stem cells. When such is the case, assumption (3) is not necessary.

3. Assumption (2) implies implicitly that the mutation process either takes place or is fixed during cell division. This is obviously the case if mutation (in the general sense) is the result of a chromosomal aberration which takes place during cell division; it is also the case for point mutation since point mutation is usually the result of error-prone repair of DNA lesion during the synthesis(S) stage of DNA (55,66,75). Note that if $c_1(t)\Delta t + o(\Delta t)$ is the probability of a cell division during $[t, t + \Delta t]$ for normal stem cells ($c_1(t) \geq \alpha_1(t) + b_{1,1}(t)$), then $\alpha_1(t)/c(t)$ is the probability that a mutation from normal stem cells takes place given a cell division of normal stem cells during $[t, t + \Delta t]$.

4. Assumption (5) is equivalent to ignoring cancer random variation in progression as defined in Chapter One. This neglect of random variation of tumor progression may have little effect on incidence curves of human cancers since life time is usually considered in human epidemiological studies. This is due to the fact that for human beings, the life span is usually very long (average 70 years) while the time for cancer progres-

sion is rather short. The fitting of the MVK two-stage model to various human cancers by Moolgavkar and Knudson (49) seems to suggest that this is in fact the case.

Some Basic Properties of the Model

Given the above two stage models and the assumptions, the following results follow immediately:

1. The process is basically Markov. This follows from the fact that any future event for any cell after time t depends only on the status of that cell at time t and is independent of past history of that cell.

2. If N_0 is very large, $b_{1j}(t) \geq d_{1j}(t)$ and if $\alpha_1(t)$ is very small, then $N(t)$ is very large for all $t \geq t_0$. Further, $N_0\alpha_1(t)$ and hence $N(t)\alpha_1(t)$ are usually finite for all $t \geq t_0$. By the limiting theory of binomial distribution, one has that to the order $0(N_0^{-1})$, the number of mutations that occur during $[t, t + \Delta t]$ from normal stem cells follows a Poisson distribution with parameter $N(t)\alpha_1(t)\Delta t + o(\Delta t)$. This result provides a justification for the assumption of Poisson distribution for the mutation of normal stem cells made in references (50) and (51) and in Theorem (3.2.3). Also, for large $N(t), t \geq t_0$, one may ignore random variation in $N(t)$ and assume some deterministic growth function for $N(t)$, an assumption made by Moolgavkar and Knudson (49) and others (44-51,68,69). This latter assumption is justifiable mathematically. In fact, if $b_{1j}(t) = jb_1(t)$ and $d_{1j}(t) = jd_1(t)$, it was shown in Tan and Brown (69) that to the order $0(N_0^{-1})$, the carcinogenesis process depends on $N(t)$ through its expected value.

3.2.2 The Kolmogorov Forward Equations and the PGF of the Number of Tumors

Let the transition probabilities be denoted by: $P_1(s,t;u_1,u_2,u_3;i_1,i_2, i_3) = P_r\{N(t) = i_1, I(t) = i_2, T(t) = i_3|N(s) = u_1, I(s) = u_2, T(s) = u_3\}$ and $P_2(s,t;v_1,v_2;j_1,j_2) = P_1\{I(t) = j_1, T(t) = j_2|I(s) = v_1, T(s) = v_2\}$.

At the starting time t_0, assume that there are no intermediate cells and no tumors. Define the conditional probabilities $P_1(i_1,i_2,i_3;t)$ and $P_2(j_1,j_2;s,t)$ by $P_1(i_1,i_2,i_3;t) = P_1(t_0,t;N_0,0,0;i_1,i_2,i_3)$ and $P_2(j_1,j_2; s,t) = P_2(s,t;1,0;j_1,j_2)$.

Let $\zeta(x_1,x_2,x_3,;t_0,t) = \zeta(t_0,t)$ be the PGF of $[N(t),I(t),T(t)]$ given $[N(t_o) = N_0, I(t_0) = T(t_0) = 0]$ and $\phi(x_2,x_3;s,t)$ the PGF of $[I(t),T(t)]$

given $[I(s) = 1, N(s) = T(s) = 0]$. That is,

$$\zeta(t_0, t) = \sum_{i_1} \sum_{i_2} \sum_{i_3} x_1^{i_1} x_2^{i_2} x_3^{i_3} \, P_1(i_1, i_2, i_3; t)$$

and

$$\phi(s, t) = \sum_{j_1} \sum_{j_2} x_2^{j_1} x_3^{j_2} P_2(j_1, j_2; s, t).$$

Obviously, $P_1(i_1, i_2, i_3; t_0) = \delta_{i_1, N_0} \delta_{0i_2} \delta_{0i_3}$ and $P_2(j_1, j_2; s, s) = \delta_{1j_1} \delta_{0j_2}$, where $\delta_{ij} = \delta_{i,j}$ is the Krenecker's δ defined by $\delta_{ij} = 1$ if $i = j$ and $\delta_{ij} = 0$ if $i \neq j$, so that $\zeta(t_0, t) = x_1^{N_0}$ and $\phi(s, s) = x_2$. Further, $\psi(t_0, t) = \psi(x_2, x_3; t_0, t) = \zeta(1, x_2, x_3; t_0, t)$ is the PGF of $[I(t), T(t)]$ given $[N(t_0) = N_0, I(t_0) = T(t_0) = 0]$ and $g(t_0, t) = g(x_3; t_0, t) = \psi(1, x_3; t_0, t)$ the PGF of $T(t)$ given $[N(t_0) = N_0, I(t_0) = T(t_0) = 0]$.

To obtain the above probabilities and the probability generating functions (PGF), a standard procedure is by way of the Kolmogorov forward equations. In this section, we proceed to develop some procedures for evaluating these PGF's. The above probabilities may then be evaluated from these PGF's as illustrated in Section (3.5).

THEOREM (3.2.1)   Define $P_1(i_1, i_2, i_3; t) = 0$ if any of (i_1, i_2, i_3) is negative and define $P_2(j_1, j_2; s, t) = 0$ if any of (j_1, j_2) is negative. Then, under assumptions (2)–(6), the $P_1(i_1, i_2, i_3; t)$'s satisfy the forward equations (3.2.1). Under assumptions (2) and (4)–(6), the $P_2(j_1, j_2; s, t)$'s satisfy equations (3.2.2).

$$\frac{d}{dt} P_1(i_1, i_2, i_3; t) = P_1(i_1 - 1, i_2, i_3; t) b_{1, i_1 - 1}(t) + P_1(i_1 + 1, i_2, i_3; t) d_{1, i_1 + 1}$$
$$+ P_1(i_1, i_2 - 1, i_3; t) i_1 \alpha_1(t) + P_1(i_1, i_2 - 1, i_3; t)(i_2 - 1) b_2(t)$$
$$+ P_1(i_1, i_2 + 1, i_3; t)(i_2 + 1) d_2(t) + P_1(i_1, i_2, i_3 - 1; t) i_2 \alpha_2(t)$$
$$- P_1(i_1, i_2, i_3; t) \{ [b_{1, i_1}(t) + d_{1, i_1}(t) + i_1 \alpha_1(t)] + i_2 [b_2(t)$$
$$+ d_2(t) + \alpha_2(t)] \}, \tag{3.2.1}$$

with $P_1(i_1, i_2, i_3; t_0) = \delta_{i_1 N_0} \delta_{0i_2} \delta_{0i_3}$, $i_1 \geq 0$, $i_2 \geq 0$, $i_3 \geq 0$;

$$\frac{d}{dt} P_2(j_1, j_2; s, t) = P_2(j_1 - 1, j_2; s, t)(j_1 - 1) b_2(t)$$
$$+ P_2(j_1 + 1, j_2; s, t)(j_1 + 1) d_2(t) + P_2(j_1, j_2 - 1; s, t) j_1 \alpha_2(t)$$
$$- P_2(j_1, j_2; s, t) j_1 [b_2(t) + d_2(t) + \alpha_2(t)], \tag{3.2.2}$$

with $P_2(j_1, j_2; s, s) = \delta_{1j_1} \delta_{0j_2}$, $j_1 \geq 0$, $j_2 \geq 0$.

REMARK Equations (3.2.1) and (3.2.2) are often referred to as the Kolmogorov forward equations. These are the basic equations by means of which the probability generating functions and hence the probabilites can be derived.

Proof. Note first that under assumptions (2)–(6), $(N(t), I(t), T(t))$ are Markov processes, while under assumptions (2) and (4)–(6), $(I(t), T(t))$ are Markov processes. Hence, by the Chapman-Kolmogorov equation,

$$P_1(i_1, i_2, i_3; t + \Delta t)$$
$$= \sum_{j_1} \sum_{j_2} \sum_{j_3} P_1(j_1, j_2, j_3; t) P_1(t, t + \Delta t; j_1, j_2, j_3; i_1, i_2, i_3).$$

Using the transition probabilities given in Table (3.1), we obtain then

$$P_1(i_1, i_2, i_3; t + \Delta t) = P_1(i_1 - 1, i_2, i_3; t) b_{1, i_1 - 1}(t) \Delta t$$
$$+ P_1(i_1 + 1, i_2, i_3; t) d_{1, i_1 + 1}(t) \Delta t + P_1(i_1, i_2 - 1, i_3; t) i_1 \alpha_1(t) \Delta t$$
$$+ P_1(i_1, i_2 - 1, i_3; t)(i_2 - 1) b_2(t) \Delta t + P_1(i_1, i_2 + 1, i_3; t)(i_2 + 1) d_2(t) \Delta t$$
$$+ P_1(i_1, i_2, i_3 - 1; t) i_2 \alpha_2(t) \Delta t + P_1(i_1, i_2, i_3; t) \{1 - [b_{1, i_1}(t)$$

Table 3.1 The Transitions and the Transition Probabilities During $[t, t + \Delta t]$ for the Two-Stage Model of Carcinogenesis

At time t	At time $t + \Delta t$	Transition Rate	Transition Probability
j N Cells ($j \geq 1$)	$j + 1$ N Cells	$b_{1j}(t)$	$b_{1j}(t) \Delta t + o(\Delta t)$
j N Cells	$j - 1$ N Cells,	$d_{1j}(t)$	$d_{1j}(t) \Delta t + o(\Delta t)$
j N Cells	j N Cells and 1 I Cell	$j \alpha_1(t)$	$j \alpha_1(t) \Delta t + o(\Delta t)$
j N Cells	j N Cells		$1 - [b_{1j}(t) + d_{1j}(t) + j \alpha_1(t)] \Delta t + o(\Delta t)$
1 I Cell	2 I Cells	$b_2(t)$	$b_2(t) \Delta t + o(\Delta t)$
1 I Cell	0 Cells	$d_2(t)$	$d_2(t) \Delta t + o(\Delta t)$
1 I Cell	1 I Cell and 1 T Cell	$\alpha_2(t)$	$\alpha_2(t) \Delta t + o(\Delta t)$
1 I Cell	1 I Cell		$1 - [b_2(t) + d_2(t) + \alpha_2(t)] \Delta t + o(\Delta t)$

$+ d_{1,i_1}(t) + i_1 \alpha_1(t)] \Delta t - i_2[b_2(t) + d_2(t) + \alpha_2(t)] \Delta t\} + o(\Delta t)$

Subtracting $P_1(i_1, i_2, i_3; t)$ on both sides, dividing by Δt and letting $\Delta t \to 0$, one obtains the forward equations for $P_1(i_1, i_2, i_3; t)$ as given in (3.2.1).

Similarly, one obtains the forward equations for $P_2(j_1, j_2; s, t)$ as given in (3.2.2).

Equations (3.2.1) and (3.2.2) provide mathematical tools for computing $P_1(i_1, i_2, i_3; t)$ and $P_2(j_1, j_2; s, t)$. To find these probabilities and to derive age specific incidence functions of tumors, we now proceed to find the PGF's $\psi(t_0, t)$ and $\phi(t_0, t)$.

THEOREM (3.2.2)   Under assumptions (2) and (4)–(6), $\psi(t_0, t)$ satisfies the first order partial differential equations (3.2.3). Under assumptions (2)–(6), $\phi(t_0, t)$ satisfies the first order partial differential equation (3.2.4).

$$\frac{\partial}{\partial t} \phi(s,t) = \{(x_2 - 1)[x_2 b_2(t) - d_2(t)]$$

$$+ x_2(x_3 - 1)\alpha_2(t)\} \frac{\partial}{\partial x_2} \phi(s,t), \phi(s,s) = x_2; \tag{3.2.3}$$

$$\frac{\partial}{\partial t} \zeta(t_0, t) = \sum_i [(x_1 - 1)b_{1,i}(t) + (x_1^{-1} - 1)d_{1,i}(t)] Q(i; t_0, t)$$

$$+ x_1(x_2 - 1)\alpha_1(t) \frac{\partial}{\partial x_1} \zeta(t_0, t) + \{(x_2 - 1)[x_2 b_2(t) - b_2(t)]$$

$$+ x_2(x_3 - 1)\alpha_2(t)\} \frac{\partial}{\partial x_2} \zeta(t_0, t), \zeta(t_0, t_0) = x_1^{N_0}, \tag{3.2.4}$$

where $Q(i; t_0, t) = x_1^i \sum_{i_2} \sum_{i_3} x_2^{i_2} x_3^{i_3} P_1(i, i_2, i_3; t)$.

Proof. To prove (3.2.3), note that

$$\sum_{j_1=0}^{\infty} \sum_{j_2=0}^{\infty} (j_1 - 1)x_2^{j_1} x_3^{j_2} P_2(j_1 - 1, j_2; s, t) =$$

$$x_2^2 \sum_{j_1=0}^{\infty} \sum_{j_2=0}^{\infty} j_1 x_2^{j_1-1} x_3^{j_2} P_2(j_1, j_2; s, t) = x_2^2 \frac{\partial}{\partial x_2} Q(s,t),$$

$$\sum_{j_1=0}^{\infty} \sum_{j_2=0}^{\infty} j_1 x_2^{j_1} x_3^{j_2} P_2(j_1, j_2; s, t) = x_2 \frac{\partial}{\partial x_2} Q(s,t),$$

$$\sum_{j_1=0}^{\infty}\sum_{j_2=0}^{\infty}j_1 x_2^{j_1}x_3^{j_2}P_2(j_1,j_2-1;s,t) =$$

$$x_2 x_3 \sum_{j_1=0}^{\infty}\sum_{j_2=0}^{\infty}j_1 x_2^{j_1-1}x_3^{j_2}P_2(j_1,j_2;s,t) = x_2 x_3 \frac{\partial}{\partial x_2}Q(s,t),$$

and

$$\sum_{j_1=0}^{\infty}\sum_{j_2=0}^{\infty}(j_1+1)x_2^{j_1}x_3^{j_2}P_2(j_1+1,j_2;s,t)$$

$$=\sum_{j_1=0}^{\infty}\sum_{j_2=0}^{\infty}j_1 x_2^{j_1-1}x_3^{j_2}P_2(j_1,j_2;s,t) = \frac{\partial}{\partial x_2}Q(s,t).$$

Then, equation (3.2.3) follows by multiplying both sides of (3.2.2) by $x_2^{j_1}x_3^{j_2}$ and summing over j_1 and j_2 from 0 to ∞.

Similarly, multiplying both sides of (3.2.1) by $x_1^{i_1}x_2^{i_2}x_3^{i_3}$ and summing over (i_1,i_2,i_3) from 0 to ∞ for each i_u one obtains equations (3.2.4) for $\zeta(t_0,t)$.

3.2.3 Some Special Cases

1. If $b_{1,i}(t) = ib_1(t)$ and $d_{1,i}(t) = id_1(t)$, then the normal stem cells follow a nonhomogeneous Feller-Arley birth-death process for their proliferation and differentiation. In these cases, equation (3.2.4) reduces to:

$$\frac{\partial}{\partial t}\zeta(t_0,t) = \{(x_1-1)[x_1 b_1(t) - d_1(t)]$$

$$+x_1(x_2-1)\alpha_1(t)\}\frac{\partial}{\partial x_1}\zeta(t_0,t)$$

$$+\{(x_2-1)[x_2 b_2(t) - d_2(t)]$$

$$+x_2(x_3-1)\alpha_2(t)\}\frac{\partial}{\partial x_2}\zeta(t_0,t), \zeta(t_0,t_0) = x_1^{N_0}. \quad (3.2.5)$$

2. If $S = \{0,1,\ldots,M\}$ and if $b_{1,i}(t) = ib_1(t)[1-(i/M)]$ and $d_{1,i}(t) = id_1(t)[1-(i/M)]$ for $i = 0,1,\ldots,M$, then the normal stem cells follow a stochastic logistic birth-death process for their proliferation and differentiation.

In these cases (3.2.4) reduces to:

$$\frac{\partial}{\partial t}\zeta(t_0,t) = \{(x_1 - 1)[x_1 b_1(t) - d_1(t)][1 - (1/M)] + x_1(x_2 - 1)\alpha_1(t)\}$$

$$\times \frac{\partial}{\partial x_1}\zeta(t_0,t) - (x_1 - 1)[x_1 b_1(t) - d_1(t)](x_1/M)\frac{\partial^2}{\partial x_1^2}\zeta(t_0,t)$$

$$+ \{(x_2 - 1)[x_2 b_2(t) - d_2(t)] + x_2(x_3 - 1)\alpha_2(t)\}\frac{\partial}{\partial x_2}\zeta(t_0,t),\ \zeta(t_0,t_0)$$

$$= x_1^{N_0}. \tag{3.2.6}$$

Equations (3.2.3), (3.2.5) and (3.2.6) are impossible to solve. However, if for all $t \geq t_0, N(t)$ is very large but $\alpha_1(t)$ very small so that $N(t)\alpha_1(t)$ is finite for all $t \geq t_0$. Then by Section(3.2.1)(b), as a close approximation one may assume a Poisson probability law for the process of mutation from normal stem cells to initiated cells. Under these conditions, $\psi(t_0,t)$ is readily available. This is illustrated in the following theorem.

THEOREM (3.2.3) Assume that $N(t)\alpha_1(t)$ is finite for all $t \geq t_0$ and that the number of mutations that occur during $[t, t + \Delta t]$ from normal stem cells follows a Poisson distribution with parameter $N(t_j)\alpha_1(t_j)\Delta t + o(\Delta t)$, independently. Then, under conditions (2) and (4)–(6), $\psi(t_0,t)$ is given by:

$$\psi(t_0,t) = exp\{\int_{t_0}^{t} N(x)\alpha_1(x)[\phi(x,t) - 1]dx\} \tag{3.2.7}$$

$$= exp\{N_0 \int_{t_0}^{t} \alpha_1(x)\mu_1(t_0,x)[\phi(x,t) - 1]dx\},$$

where $\mu_1(t_0,x)$ is the expected number of normal stem cell at time x given one normal stem cell at t_0.

Proof. Partition the time interval $[t_0,t]$ by $L_j = [t_j, t_{j+1}), j = 0, 1,\ldots,n-1$, where $t_j = t_0 + j\Delta t$ and $n\Delta t = t - t_0 = t_n - t_0$. Let M_j be the number of initiated cells arising from normal stem cells during $L_j, j = 0, 1,\ldots,n-1$. Then the conditional PGF of $I(t)$ and $T(t)$ given the partition and given $(M_j, j = 0, 1,\ldots,n - 1)$ is :

$$\xi(t;\Delta t) = \prod_{j=0}^{n-1}[\phi(t_j,t)]^{M_j}.$$

Since M_j follows a Poisson distribution with parameter $N(t_j)\alpha_1(t_j)\Delta t + o(\Delta t)$ independently for $j = 0, 1, \ldots, n-1$, so, by taking expectation over $[M_j, j = 0, 1, \ldots, n-1]$, one obtains the conditional PGF of $I(t)$ and $T(t)$ given the partition and given $[N(t_0) = N_0, I(t_0) = T(t_0) = 0]$ as

$$Q_p(t_o, t; \Delta t) = exp\{\sum_{j=0}^{n-1} N(t_j)\alpha_1(t_j)[\phi(t_j, t) - 1]\Delta t\} + (t - t_0)o(\Delta t)/\Delta t.$$

It follows that

$$\psi(t_0, t) = \lim_{\Delta t \to 0} Q_p(t_0, t; \Delta t) = exp\{\int_{t_0}^{t} N(u)\alpha_1(u)[\phi(u, t) - 1]du\}$$

In (3.2.7) of Theorem (3.2.3), $N(x)$ is the number of normal stem cells at time x given $N(t_0) = N_0$ ($x \geq t_0$). Since $N(t)$ is very large for all $t \geq t_0$, one may take $N(x)$ as the expected number of normal stem cells at time x given $N(t_0) = N_0$. $N(x)$ in general depends on the growth pattern of the tissue and hence $N(x)$ is the major factor determining the shape of incidence functions of cancers for different tissues.

Tan and Brown (69) have shown that if $b_{1j}(t) = jb_1(t)$ and $d_{ij}(t) = jd_1(t)$, then the above two-stage model is a continuous time multiple branching process. Thus, an alternative approach to prove formula (3.2.7) is by way of a continuous time multiple branching process. In this approach, the Poisson distribution assumption for the mutation of normal stem cells is replaced by assuming $b_{1j}(t) = jb_1(t)$ and $d_{1j}(t) = jd_1(t)$ in assumption (3); for details, see reference (69).

3.3 SOME SPECIAL CASES OF TWO-STAGE MODELS

The two-stage models described in previous sections are quite nonspecific for general nonhomogeneous cases. In this general setup, the solution of $\phi(x, t)$ in equation (3.2.7) is very difficult to derive. To obtain $\psi(t_0, t)$, we thus consider some important special cases. The following three models cover most of the cases in application.

3.3.1 The Moolgavkar-Venzon-Knudson Two-Stage Model

The two-stage models as proposed by Moolgavkar and Venzon (47) and Moolgavkar and Knudson (49) deal only with homogeneous cases. (This model will be referred to as the MVK model.) In this model, $\alpha_i(t) = \alpha_i, b_2(t) = b_2, d_2(t) = d_2$ and $N(t)$ are deterministic growth curve functions. In this case, $\phi(t_0,t) = \phi(t - t_0)$ is a function of $t - t_0$. Putting $t_0 = 0$, then the calender time may be taken as a person's age.

THEOREM (3.3.1) If $\alpha_2(t) = \alpha_2, b_2(t) = b_2$ and $d_2(t) = d_2$, then $\phi(x_2,x_3;0,t) = \phi(t)$ satisfies the following Ricatti equation with initial condition $\phi(0) = x_2$:

$$\frac{d}{dt}\phi(t) = b_2\phi^2(t) + [\alpha_2 x_3 - (b_2 + d_2 + \alpha_2)]\phi(t) + d_2, \qquad (3.3.1)$$

If further the normal stem cells follow a homogeneous Feller-Arley birth-death process with birth rate b_1 and death rate d_1, then the PGF $\psi_1(x_1,x_2,x_3;0,t) = \psi_1(t)$ of $[N(t),I(t),T(t)]$ given $N(0) = 1$ satisfies the following Ricatti equation with initial condition $\psi_1(0) = x_1$:

$$\frac{d}{dt}\psi_1(t) = b_1\psi_1^2(t) + [\alpha_1\phi(t) - (b_1 + d_1 + \alpha_1)]\psi_1(t) + d_1. \qquad (3.3.2)$$

Proof. We shall prove (3.3.1) only as the proof of (3.3.2) is exactly the same.

Starting with one initiated cell at time 0, the end result at time t depends on events happening during $[0,t)$. These events are either (i) no change during $[0,t)$ or (ii) a change happened during $[0,t)$. If a change takes place during $[0,t)$, it is then either a birth, or a death or a mutation to yield a tumor cell.

To obtain the probability of no change during $[0,t]$, we partition $[0,t)$ by $L_j = [t_j, t_{j+1})$, where $t_j = j\Delta t$ and $t = t_n = n\Delta t, j = 0,1,\ldots,n-1$. Then the probability of no change during $L_j = [t_j, t_{j+1}) = [j\Delta t, (j+1)\Delta t)$ is $[1 - (b_1 + d_1 + \alpha_1)\Delta t] + o(\Delta t)$. It follows that the probability of no change during $[0,t)$ is

$$\lim_{\Delta t \to 0}\{[1 - (b_1 + d_1 + \alpha_1)\Delta t] + o(\Delta t)\}^n$$

$$= \lim_{\Delta t \to 0}\exp\{n\log[1 - (b_1 + d_1 + \alpha_1)\Delta t]\}$$

$$= \exp\{-(b_1 + d_1 + \alpha_1)t\}, \quad (\text{note that } n\Delta t = t).$$

Since with probability one each tumor cell develops instantaneously into a tumor, the above arguments then lead to:

$$\phi(t) = x_2 \exp(-C_2 t) + \int_0^t \exp(-C_2 y)[b_2 \phi^2(t-y) + \alpha_2 \phi(t-y)x_3$$

$$+d_2]dy = x_2 \exp(-C_2 t) + \int_o^t \exp[-(t-y)C_2][b_2 \phi^2(y)$$

$$+\alpha_2 x_3 \phi(y) + d_2]dy, \text{ where } C_2 = b_2 + d_2 + \alpha_2.$$

Multiplying both sides of the above equation by $\exp(C_2 t)$ gives:

$$exp(C_2 t)\phi(t) = x_2 + \int_0^t \exp(C_2 y)[b_2 \phi^2(y) + \alpha_2 x_3 \phi(y) + d_2]dy.$$

Equation (3.2.1) follows by taking the derivative on both sides of the equation and then multiplying both sides by $exp\ (-C_2 t)$.

Equations (3.3.1) and (3.3.2) are the basic results used by Moolgavkar and Venzon (47) to derive incidence functions. Solution of equation (3.3.2) is very difficult since it is a nonlinear Ricatti equation. Although Moolgavkar and Venzon (47) gave an infinite series solution to equation (3.3.2), however, they used instead numerical solutions of equation (3.3.2) to compute some incidence curves due presumably to the slow convergence of the series. On the other hand, equation (3.3.1) can readily be solved to give the solution as:

$$\phi(t) = \{y_2(x_2 - y_1) + y_1(y_2 - x_2)exp[b_2(y_2 - y_1)t]\}$$
$$\times \{(x_2 - y_1) + (y_2 - x_2)exp[b_2(y_2 - y_1)t]\}^{-1} \tag{3.3.3}$$

where $y_2 > y_1$ are given by $2b_2 y_2 = (b_2 + d_2 + \alpha_2 - \alpha_2 x_3) + h(x_3)$ and $2b_2 y_1 = (b_2 + d_2 + \alpha_2 - \alpha_2 x_3) - h(x_3)$ with $h(x_3) = [(b_2 + d_2 + \alpha_2 - \alpha_2 x_3)^2 - 4b_2 d_2]^{\frac{1}{2}}$. (Note $h(x_3) \geq 0$ so that $y_2 > y_1$ are real numbers for all x_3.) For the details of solving equation (3.3.1), see Section (2.5.2) of Chaper Two. Note that as shown in Section (2.5.1) of Chapter Two, (3.3.3) is also the solution of equation (3.3.3) in the homogeneous cases.

3.3.2 The Model with $d_2(t) = 0$

Mackillop et al. (38), Buick and Pollak (9), Oberley and Oberley (53) have provided much evidence supporting the idea that normal stem cells become immortalized by the loss of differentiation capability (see also

41,42). These results suggest that one may assume $d_2(t) = 0$ to obtain $\psi(t_0,t)$. In this case, equation (3.2.3) reduces to:

$$\frac{\partial}{\partial t}\phi(t_0,t) = x_2[(x_2 - 1)b_2(t) + (x_3 - 1)\alpha_2(t)]\frac{\partial}{\partial x_2}\phi(t_0,t), \phi(t_0,t_0) = x_2.$$

This equation has been solved in Section (2.5.1) of Chapter Two to yield the solution as

$$\phi(t_0,t) = x_2/\{\xi(x_3;t_0,t) - x_2\eta(x_3;t_0,t)\}, \tag{3.3.4}$$

where $\xi(x_3;t_0,t) = \exp\{\int_{t_0}^{t}[b_2(x) + (1 - x_3)\alpha_2(x)]dx\}$ and $\eta(x_3;t_0,t) = \int_{t_0}^{t} b_2(u)\xi(x_3;t_0,u)du$.

3.3.3   The Tan-Gastardo Two-Stage Model

In many cases, it appears that the life time can be partitioned into several time intervals in each of which one may assume a homogeneous model. For example, for female breast and ovary cancers in women, menarche, first pregnancy and menopause provide natural partition of the life time intervals. In initiation and promotion experiments with animals, during certain time intervals, the animals (rats for example) are exposed to some initiators while in some latter time intervals, the animals are exposed to promoters. For these experiments, to assess effects of environmental agents, it is natural to consider a model in which the life time interval is divided into several nonoverlapping time intervals in each of which one may assume a homogeneous model. For assessing effects of environmental agents from these experiments, Tan and Gastardo (68) have extended the Moolgavkar and Knudson (49) homogeneous two stage models into a nonhomogeneous model involving several time intervals.

To illustrate the model, let the time interval $[t_0,t]$ be partitioned into k sub-intervals $L_j = [t_{j-1},t_j), j = 1,\ldots,k - 1$ and $L_k = [t_{k-1},t_k]$ with $t = t_k$ and assume that in the jth interval $L_j, \alpha_2(t) = \alpha_{2j}, b_2(t) = b_{2j}$ and $d_2(t) = d_{2j}, j = 1,\ldots,k$.

Let $y_2 > y_1$ be the two real roots of $bx^2 - (b + d + \alpha - \alpha x_3)x + d = 0$ and define $\phi(x_2,x_3;t,b,d,\alpha)$ by:

$$\phi(x_2,x_3;t,b,d,\alpha) = \{y_2(x_2 - y_1) + y_1(y_2 - x_2)\exp[b(y_2 - y_1)t]\}$$
$$\times\{(x_2 - y_1) + (y_2 - x_2)\exp[b(y_2 - y_1)t]\}^{-1} \tag{3.3.5}$$

Then as shown in Section (2.5.1) of Chapter Two and Section (3.3.1) above, $\phi(x_2,x_3;t,b_{2k},d_{2k},\alpha_{2k})$ is the solution of the following equation with initial condition $g_k(0) = x_2$:

$$\frac{\partial}{\partial t} g_k(t) = [(x_2 - 1)(x_2 b_{2k} - d_{2k}) + x_2(x_3 - 1)\alpha_{2k}] \frac{\partial}{\partial x_2} g_k(t). \quad (3.3.6)$$

To obtain solution (3.2.5), define for $t_{j-1} \leq u < t_j, j = 1,\ldots,k-1$:

$$g_j(u,t_k) = \phi\{g_{j+1}(t_j,t_k), x_3; t_j - u, b_{2j}, d_{2j}, \alpha_{2j}\} \text{ with } g_k(t_{k-1},t_k) = g_k(\tau_k), \quad (3.3.7)$$

where $\quad \tau_j = t_j - t_{j-1}, j = 1,\ldots,k$.

Then, by the chain rule, one has, for $t = t_k$ and $t_{j-1} \leq u < t_j, j = 1,\ldots,k-1$:

$$\begin{aligned}
\frac{\partial}{\partial t} g_j(u,t) &= \frac{\partial}{\partial t} \phi\{g_{j+1}(t_j,t), x_3; t_j - u, b_{2j}, d_{2j}, \alpha_{2j}\} \\
&= \{\partial\phi(y,x_3; t_j - u, b_{2j}, d_{2j}, \alpha_{2j})/\partial y\}_{[y=g_{j+1}(t_j,t)]} \\
&\quad \times \{\partial g_{j+1}(t_j,t)/\partial t\} \\
&= \{\partial\phi(y,x_3; t_j - u, b_{2j}, d_{2j}, \alpha_{2j})/\partial y\}_{[y=g_{j+1}(t_j,t)]} \\
&\quad \times \prod_{v=j+1}^{k-1} \{\partial\phi(y,x_3; \tau_v, b_{2v}, d_{2v}, \alpha_{2v})/\partial y\}_{[y=g_{v+1}(t_v,t)]} \\
&\quad \times \{\partial g_k(\tau_k)/\partial\tau_k\} \quad (3.3.8)
\end{aligned}$$

Similarly, by the chain rule, for $t = t_k$ and $t_{j-1} \leq u < t_j, j+1,\ldots,k$:

$$\begin{aligned}
\frac{\partial}{\partial x_2} g_j(u,t) &= \{\partial\phi(y,x_3; t_j - u, b_{2j}, d_{2j}, \alpha_{2j})/\partial y\}_{[y=g_{j+1}(t_j,t)]} \\
&\quad \times \prod_{v=j+1}^{k-1} \{\partial\phi(y,x_3; \tau_v, b_{2v}, d_{2v}, \alpha_{2v})/\partial y\}_{[y=g_{v+1}(t_v,t)]} \{\partial g_k(\tau_k)/\partial x_2\} \quad (3.3.9)
\end{aligned}$$

Combining (3.3.8) and (3.3.9) with (3.3.6), it is clear that $g_j(u,t)$ satisfies the following equation with initial condition $g_j(u,u) = x_2$,

$$\frac{\partial}{\partial t} g_j(u,t) = \{(x_2 - 1)[x_2 b_2(t) - d_2(t)] + x_2(x_3 - 1)\alpha(t)\} \frac{\partial}{\partial x_2} g_j(u,t) \quad (3.3.10)$$

for all $t_{j-1} \leq u < t_j, j = 1,\ldots,k$. (Note that with $t = t_k, b_2(t) = b_{2k}, d_2(t) = d_{2k}$ and $\alpha_2(t) = \alpha_{2k}$; further, $t = u$ implies $\tau_v = 0$ for all $v = j+1,\ldots,k$.)
Note that (3.3.10) is exactly the same as (3.2.3).

On substituting this result into $\psi(t_0, t)$, one has with $t = t_k$:

$$\psi(t_0, t) = \exp \{N_0 \int_{t_0}^{t} \alpha_1(u)\mu_1(t_0, u)[\phi(u, t) - 1]du\}$$

$$= \exp \{N_0 \sum_{j=1}^{k} \int_{t_{j-1}}^{t_j} \alpha_1(u)\mu_1(t_0, u)[\phi(u, t) - 1]du\}$$

$$= \exp \{\sum_{j=1}^{k} N_{j-1} \int_{t_{j-1}}^{t_j} \alpha_1(u)\mu_1(t_{j-1}, u)[g_j(u, t) - 1]du\} \quad (3.3.11)$$

where $g_j(u, t)$ is defined by (3.3.7), and N_{j-1} is the expected number of normal stem cells at $t_{j-1}, i.e. N(t_{j-1}) = N_{j-1}$.

Formula (3.3.11) was first obtained by Tan and Gastardo (68). In (3.3.11), observe that $g_j(u, t)$ is the PGF of $I(t)$ and $T(t)$ given one initiated cell at time u, where $t_{j-1} \leq u < t_j$.

3.4 THE AGE-SPECIFIC INCIDENCE FUNCTIONS OF TUMORS

Let T be the time to tumor starting with N_0 normal stem cells at t_0 and let $\lambda(t)$ be the incidence function of tumors given N_0 normal stem cells at t_0. Then, the probability density function of T is given by:

$$f_T(t) = \lambda(t)exp \{-\int_{t_0}^{t} \lambda(x)dx\}, t \geq t_0,$$

$$= 0, \quad \text{otherwise.}$$

Hence, if $\lambda(t)$ is available, one may obtain $f_T(t)$. In this section I proceed to obtain $\lambda(t)$ for the above two-stage model of carcinogenesis and some important special cases.

3.4.1 Incidence Functions for General Cases and an Approximation

By Theorem (2.1.1) of Chapter Two,

$$\lambda(t) = -g'(0; t_0, t)/g(0; t_1, t) = -\psi'(1, 0; t_0, t)/\psi(1, 0; t_0, t)$$

where

$$g'(0, t_0, t) = dg(0; t_0, t)/dt = d\psi(1, 0; t_0, t)/dt = \psi'(1, 0; t_0, t).$$

From (3.2.7) of Section (3.2),

$$\psi'(1,0;t_0,t) = \psi(1,0;t_0,t)\frac{d}{dt}\int_{t_0}^{t} N(x)\alpha_1(x)[\phi(1,0;x,t)-1]dx\}$$

$$= \psi(1,0;t_0,t)\{N(t)\alpha_1(t)[\phi(1,0;t,t)-1]$$

$$+ \int_{t_0}^{t} N(x)\alpha_1(x)[\partial\phi(z,0;x,t)/\partial t]dx\}$$

Since $\phi(x_2,x_3;t,t) = x_2$, so

$$\lambda(t) = -\psi'(1,0;t_0,t)/\psi(1,0;t_0,t)$$

$$= -\int_{t_0}^{t} N(x)\alpha_1(x)[\partial\phi(1,0;x,t)/\partial t]dx.$$

From equation (3.2.3) of Section (3.2),

$$\partial\phi(1,0;x,t)/\partial t = [\partial\phi(y,z;x,t)/\partial t]_{(y=1,z=0)} = -\alpha_2(t)u_{2,0}(x,t),$$

where $u_{2,0}(x,t) = [\partial\phi(y,0;x,t)/\partial y]_{(y=1)}$. Hence,

$$\lambda(t) = \alpha_2(t)\int_{t_0}^{t} N(x)\alpha_1(x)u_{2,0}(x,t)dx$$

$$= N_0\alpha_2(t)\int_{t_0}^{t} \alpha_1(x)\mu_1(t_0,x)u_{2,0}(x,t)dx \qquad (3.4.1)$$

From (3.4.1), to obtain $\lambda(t)$ by using formula (3.4.1) one would need the solution $\phi(x_2,0;t_0,t)$ of (3.2.3). As shown in the previous section, for some important special cases, $\phi(y,0;t_0,t)$ is readily available.

An Approximation for $\lambda(t)$

Since $\phi(x_2,0;s,t)$ is not available for general cases, one would need to find approximations for $\lambda(t)$ in general cases. Now, $\epsilon_2(t) = b_2(t)-d_2(t)$ is usually not big ($\epsilon_2(t) \leq 0.1$) and $\alpha_2(t)$ is usually very small ($\alpha_2(t) \approx 10^{-6}-10^{-8}$) in most of the cases. A close approximation to $\lambda(t)$ can readily be derived. To obtain such an approximation, observe that if $\alpha_2(t) = 0$ for all $t \geq t_0$, then equation (3.2.3) reduces to:

$$\frac{\partial}{\partial t}\phi(s,t) = (x_2-1)[(x_2-1)b_2(t)+\epsilon_2(t)]\frac{\partial}{\partial x_2}\phi(s,t), \phi(s,s) = x_2. \quad (3.4.2)$$

Equation (3.4.2) has been solved in Section (2.5.1) of Chapter Two to give:

$$\phi(s,t) = 1 - (x_2 - 1)\{(x_2 - 1)\eta_0(s,t) - \xi_0(s,t)\}^{-1}, \qquad (3.4.3)$$

where $\xi_0(s,t) = exp[-\int_s^t \epsilon_2(x)dx]$ and $\eta_0(s,t) = \int_s^t b_2(u)\xi_0(s,u)du$.

From (3.4.3), if $\alpha_2(t) \approx 0$ and if $\epsilon_2(t)$ is small for all $t \geq t_0$, a close approximation to $\phi(y,0;x,t)$ is given by

$$\phi(y,0;x,t) \approx 1 - (y - 1)\{(y - 1)\eta_0(x,t) - \xi_0(x,t)\}^{-1}.$$

Thus, $\partial\phi(y,0;x,t)/\partial y \cong \xi_0(x,t)\{(y - 1)\eta_0(x,t) - \xi_0(x,t)\}^{-2}$ so that if $\alpha_2(t) \approx 0$, and if $\epsilon_2(t)$ is small,

$$[\partial\phi(y,0;x,t)/\partial y]_{y=1} \approx exp[\int_x^t \epsilon_2(u)du].$$

It follows that if $\alpha_2(t) \cong 0$, and if $\epsilon_2(t)$ is small, a close approximation to $\lambda(t)$ is

$$\lambda(t) \cong N_0\alpha_2(t)\int_{t_0}^t \alpha_1(x)\mu_1(t_0,x)exp[\int_x^t \epsilon_2(u)du]dx, \qquad (3.4.4)$$

Tan and Brown (69), Thorslund, Brown and Charnley (74) have used formula (3.4.4) to obtain $\lambda(t)$. Moolgavkar, Dewanji and Venzon (50) noted that unless $\alpha_2(t)$ and/or $\epsilon_2(t)$ is large, (3.4.4) provides a close approximation to $\lambda(t)$; however, if $\alpha_2(t)$ and/or $\epsilon_2(t)$ is very large, the approximation (3.4.4) may deviate significantly from the true values.

3.4.2 Some Important Special Cases

The Moolgavkar-Venzon-Knudson Model

If $\alpha_2(t) = \alpha_2, b_2(t) = b_2$ and $d_2(t) = d_2$ are independent of t, then $\phi(y,0;t_0,t) = \phi(y,0;t-t_0)$ is available from Section (3.3.1) of this chapter.

To obtain $\partial\phi(y,0;t - t_0)/\partial y$, put $h(\alpha_2) = [(b_2 + d_2 + \alpha_2)^2 - 4b_2d_2]^{1/2}$ and let $y_2 > y_1$ be given by: $2b_2y_2 = (b_2 + d_2 + \alpha_2) + h(\alpha_2)$ and $2b_2y_1 = (b_2 + d_2 + \alpha_2) - h(\alpha_2)$. Then, $\phi(y,0;t - t_0)$ is given by:

$$\phi(y,0;t - t_0) = \{y_2(y - y_1) + y_1(y_2 - y)exp[h(\alpha_2)(t - t_0)]\}$$
$$\times \{(y - y_1) + (y_2 - y)exp[h(\alpha_2)(t_0)]\}^{-1}, \qquad (3.4.5)$$

Taking the derivative of (3.4.5) with respect to y, one obtains:

$$[\partial\psi(y,0;t - t_0)/\partial y]_{(y=1)} = (y_2 - y_1)^2exp[h(\alpha_2)(t - t_0)]$$
$$\times \{(1 - y_1) + (y_2 - 1)exp[h(\alpha_2)(t - t_0)]\}^{-2}$$

If $\alpha_2(t) = 0$, then $h(0) = b_2 - d_2, y_2 = 1$ and $y_1 = d_2/b_2$, so that

$$[\partial\phi(y,0;t-t_0)/\partial y]_{(y=1)} = exp[(b_2 - d_2)(t - t_0)].$$

Thus, if $\alpha_2(t) \approx 0$, and if $\epsilon_2 = b_2 - d_2$ is small, $\lambda(t)$ is approximated by

$$\lambda(t) \approx N_0\alpha_2 \int_{t_0}^{t} \alpha_1(x)\mu_1(t_0,x)exp[(b_2 - d_2)(t - x)]dx. \qquad (3.4.6)$$

This formula was first given by Moolgavkar and Knudson (49) who have also assumed $\alpha_1(t) = \alpha_1$.

Alternatively, one may also obtain $u_{2,0}(t_0,t) = u_{2,0}(t-t_0) = [\partial\phi(y,0;t-t_0)/\partial y]_{(y=1)}$ from the Ricatti equation:

$$\frac{d}{dt}\phi(y,0;t) = b_2\phi^2(y,0;t) - (b_2 + d_2 + \alpha_2)\phi(y,0;t) + d_2, \phi(y,0;0) = y.$$

(See Section (3.3.1) of this chapter).

Taking the derivative with respect to y on both sides and setting $y = 1$ gives:

$$\frac{d}{dt}u_{2,0}(t) = 2b_2\phi_0(t)u_{2,0}(t) - (b_2 + d_2 + \alpha_2)u_{2,0}(t), \quad \text{with} \quad u_{2,0}(0) = 1,$$

where $\phi_0(t) = \phi(1,0;t)$.

Hence

$$u_{2,0}(t) = [\phi(y,0;t)/\partial y]_{(y=1)} = exp[2b_2\int_0^t \phi_0(u)du - (b_2 + d_2 + \alpha_2)t],$$

so that

$$\lambda(t) = N_0\alpha_2 \int_{t_0}^{t} \alpha_1(x)\mu_1(t_0,x)exp[-(b_2 + d_2 + \alpha_2)(t - x)$$

$$+ 2b_2 \int_0^{t-x} \phi_0(u)du]dx, \qquad (3.4.7)$$

If $\alpha_2 = 0$, then $y_2 = 1$ and $y_1 = d_2/b_2$ so that $\phi_0(t) = \phi(1,0;t) = 1$. Thus, if $\alpha_2 \approx 0$ and if ϵ_2 is small, (3.4.7) reduces to (3.4.6).

Assuming $\alpha_1(t) = \alpha_1$, Moolgavkar and Knudson (49) used formula (3.4.6) to fit incidence curves of various human cancers. The shape of the incidence curves appears to be determined by $N_0\mu_1(t_0,t) = N(t)$, the expected number of normal stem cells at time t given N_0 normal stem cells at time t_0. The functional form of $N(t)$ for different tissues are specified by the growth pattern of the tissues. We shall illustrate this in Section (3.4.3) of this chapter.

The Model with $d_2(t) = 0$

As we have emphasized in the previous section, much evidence exists indicating that normal stem cells become immortalized by the loss of differentiation capability (9,38,53). These results suggest that one may perhaps assume $d_2(t) = 0$. In this case, then $\phi(y, 0; t_0, t)$ is available from (3.3.4) of Section (3.3.2) of this chapter. In fact, one has:

$$\phi(y, 0; t_0, t) = y / \{\xi(0; t_0, t) - y\eta(t_0, t)\},$$

where $\xi(0; t_0, t) = \exp\{\int_{t_0}^{t} [b_2(x) + \alpha_2(x)]dx\} = \xi(t_0, t)$

and $\eta(t_0, t) = \int_{t_0}^{t} b_2(u)\xi(t_0, u)du.$ (3.4.8)

Hence, one obtains from (3.4.8),

$$[\partial\phi(y, 0; t_0, t)/\partial y]_{(y=1)} = \xi(t_0, t)\{\xi(t_0, t) - \eta(t_0, t)\}^{-2}.$$

Thus,

$$\lambda(t) = \alpha_2(t) \int_{t_0}^{t} N_0\alpha_1(x)\mu_1(t_0, x)\xi(x, t)[\xi(x, t) - \eta(x, t)]^{-2}dx, \quad (3.4.9)$$

If $\alpha_2(t) = \alpha_2$ and $b_2(t) = b_2$, then $\xi(x, t) - \eta(x, t) = \exp[(b_2 + d_2)(t - x)] - [b_2/(b_2 + \alpha_2)]\{\exp[(b_2 + \alpha_2)(t - x)] - 1\} = (b_2 + \alpha_2)^{-1}\{b_2 + \alpha_2 \exp[(b_2 + \alpha_2)(t - x)]\}$ so that (3.4.9) reduces to

$$\lambda(t) = \alpha_2(b_2 + \alpha_2)^2 \int_{t_0}^{t} N_0\alpha_1(x)\mu_1(t_0, x)$$

$$\times \exp[(b_2 + \alpha_2)(t - x)]\{b_2 + \alpha_2 \exp[(b_2 + \alpha_2)(t - x)]\}^{-2}dx, \quad (3.4.10)$$

In $\alpha_2 \cong 0$ and if b_2 is small, (3.4.10) is then approximated by:

$$\lambda(t) \cong \alpha_2(t) \int_{t_0}^{t} N_0\alpha_1(x)\mu_1(t_0, x) \exp[b_2(t - x)]dx.$$

The Tan-Gastardo Model

In this model, the life time $[t_0, t]$ is divided into k nonoverlapping intervals $L_j = [t_{j-1}, t_j), j = 1, \ldots, k - 1$ and $L_k = [t_{k-1}, t_k], t = t_k$ such that in $L_j, b_2(t) = b_{2j}, d_2(t) = d_{2j}$ and $\alpha_2(t) = \alpha_{2j}$.

Writing $g_j(u,t)$ in (3.3.6) of Section (3.3.3) as $g_j(u,t) = g_j(x_2,x_3;u,t)$, then from formula (3.3.11) of Section (3.3.3), one has:

$$\psi(y,0;t_0,t) = \exp\{N_0 \int_{t_0}^{t} \alpha_1(x)\mu_1(t_0,x)[\phi(y,0;x,t)-1]dx\}$$

$$= \exp\{\sum_{j=1}^{k} N_{j-1} \int_{t_{j-1}}^{t_j} \alpha_1(x)\mu_1(t_{j-1},x)[g_j(y,0;x,t)-1]dx\}, t = t_k.$$

Thus

$$\lambda(t) = -\psi'(1,0;t_0,t)/\psi(1,0;t_0,t) \tag{3.4.11}$$

$$= \alpha_2(t)\sum_{j=1}^{k} N_{j-1} \int_{t_{j-1}}^{t_j} \alpha_1(x)\mu_1(t_{j-1},x)[\partial g_j(y,0;x,t)/\partial y]_{(y=1)}dx$$

To obtain $[\partial g_j(y,0;x,t)/\partial y]_{(y=1)}$, note from Section (3.3.3) of this chapter that $g_k(y,0;s,t) = \phi(y,0;t-s,b_{2k},d_{2k},\alpha_{2k})$ which is defined in formula (3.3.5) and for $t_{j-1} \le s < t_j < t = t_k, j = 1,\ldots,k-1, g_j(y,0;s,t) = \phi\{g_{j+1}(y,0;t_j,t),x_3;t_j-s,b_{2j},d_{2j},\alpha_{2j}\}$ with $g_k(y,0;t_{k-1},t_k) = g_k(y,0;t_k-t_{k-1})$ as defined in formular (3.3.7). Put $g_k(y,0;s,t) = h_k(y;t-s)$ and $h_k(1;t-s) = h_{k,0}(t-s)$; and for $t_{j-1} \le s < t_j < t, j = 1,\ldots,k-1$, put $g_j(y,0;s,t) = h_j(y;s,t)$ and $h_j(1;s,t) = h_{j,0}(t_j-s)$. Then one has:

THEOREM (3.4.1). Define

$$\sum_{j=k+1}^{k} \text{ as zero and } \prod_{j=k+1}^{k} \text{ as 1.}$$

Then, for $j = 1,2,\ldots,k$ and for fixed $t = t_k$,

$$[\partial g_j(y,0;u,t)/\partial y]_{(y=1)} = m_j(t_j-u) \prod_{v=j+1}^{k} m_v(\tau_v), \tag{3.4.12}$$

where $\tau_v = t_v - t_{v-1}$, and where

$$m_v(x) = \exp\{2b_v \int_0^x h_{v,0}(y)dy - (b_{2v} + d_{2v} + \alpha_{2v})x\}, v = 1,\ldots,k.$$

Proof. By the results of Section (3.4.2)(a), $[\partial h_k(y;u,t)/\partial y]_{(y=1)} = m_k(t-u)$. Let $u_j(x) = [\partial\phi(z,0;x;b_{2j},d_{2j},\alpha_{2j})/\partial z]_{[z=h_{j+1}(1;t_j,t)]}$. By the chain rule,

$$[\partial h_{k-1}(y;v,t_k)/\partial y]_{(y=1)}$$

$$= \{\partial\phi[h_k(y;t_{k-1},t_k),0;t_{k-1}-v,b_{2,k-1},d_{2,k-1},\alpha_{2,k-1}]/\partial y\}_{(y=1)}$$

$$= u_{k-1}(t_{k-1}-v)[\partial h_k(y;t_{k-1},t_k)/\partial y]_{(y=1)} = u_{k-1}(t_{k-1}-v)m_k(\tau_k).$$

By mathematical induction,

$$[\partial h_j(y;v,t_k)/\partial y]_{y=1} = u_j(t_j-v)\prod_{l=j+1}^{k-1}u_l(\tau_l)m_k(\tau_k), \tag{3.4.13}$$

Now, with $\phi_j(x) = \phi(z,0;x;b_{2j},d_{2j},\alpha_{2j})$, $\phi_j(0) = z$ (see (3.3.5) of Section (3.3.3)), and $h_{j,0}(x) = [\phi_j(x)]_{[z=h_{j+1}(1;t_j,t)]}$; further, $\phi_j(x)$ satisfies the following Ricatti equation:

$$\frac{d}{dx}\phi_j(x) = b_{2j}\phi_j^2(x) - (b_{2j}+d_{2j}+\alpha_{2j})\phi_j(x) + d_{2j}.$$

(See (3.3.1) of Section (3.3.1)).

Taking derivatives on both sides with respect to z and setting $z = h_{j+1}(1;t_j,t)$, one obtains:

$$\frac{d}{dx}u_j(x) = 2b_{2j}h_{j,0}(x)u_j(x) - (b_{2j}+d_{2j}+\alpha_{2j})u_j(x), \tag{3.4.14}$$

To obtain an initial condition for (3.4.14), observe that $\phi_j(0) = z$ so that $\partial\phi_j(0)/\partial z = 1$ and $u_j(0) = 1$.

The solution of equation (3.4.14) with $u_j(0) = 1$ is

$$u_j(x) = \exp\left\{2b_{2j}\int_0^x h_{j,0}(u)du - (b_{2j}+d_{2j}+\alpha_{2j})x\right\} = m_j(x).$$

On substituting $u_j(x) = m_j(x)$ into (3.4.13) with $v_j = t_j - v$, (3.4.12) of Theorem (3.4.1) follows.

Substituting (3.4.12) into $\lambda(t)$, one obtains:

$$\lambda(t) = \alpha_2(t)\sum_{j=1}^{k}N_{j-1}\int_{t_{j-1}}^{t_j}\alpha_1(x)\mu_1(t_{j-1},x)m_j(t_j-x)dx\prod_{u=j+1}^{k}m_u(\tau_u). \tag{3.4.15}$$

Formula (3.4.15) was first obtained by Tan and Gastardo (68) by using a different approach. Note that if $\alpha_2(t)$ is very small, then $h_{j,0}(u) \cong 1$ (see Section (3.4.2)(c)), so that $m_j(x) \cong \exp[(b_{2j} - d_{2j} - \alpha_{2j})x]$. Using (3.4.15), Tan and Gastardo (68) have computed some incidence functions for some hypothetical initiation and promotion experiments. Given in Table (3.2) are incidence rates for some computer simulated initiation and promotion experiments. In these computer simulated experiments, the time interval $[0,t]$ with $t \geq 65$, was divided into five subintervals with length (10, 20, 15, 10, t-55) units. The initiator was applied at the second time intervals with concentrations (denoted by C_{Ij}) 5, 25, 50, 100; the promoter was applied in the third time interval and fourth time interval with concentrations (denoted by C_{pj}) 10, 20, 30. The parameter values are taken as $N_0 = 10^6, b_{2j} = 0.04 + C_{pj}(0.001)$, $d_{2j} = 0.04, \alpha_{1j} = 0.3 \times 10^{-6} + C_{Ij}(1.563 \times 10^{-5}), \alpha_{2j} = 0.3 \times 10^{-6}, C_{Ij}$ and C_{pj} being the concentrations of initiator and promoter at the jth time interval respectively. The growth of normal stem cells are assumed to follow a homogeneous Feller-Arley birth-death process with birth rate $b_1 = 0.04$ and death rate $d_1 = 0.004$.

3.4.3 Incidence Functions of Human Cancer

One of the major differences between the MVK two-stage model and the two-stage models of Armitage and Doll (2), Kendall (28) and Neyman and Scott (52) is that Moolgavkar and Knudson (49) took into account different growth patterns of different tissues. This is perhaps the main

Table 3.2 Incidence Rate of Tumors in a Computer Simulated Initiation-Promotion Experiment. (Parameters are Specified in Section (3.4.2)(c)).

		t_k						
C_I	C_P	65	70	75	80	85	90	95
5	10	.00622	.00747	.00898	.01078	.01294	.01552	.01862
	20	.00795	.00954	.01145	.01374	.01648	.01977	.02370
	30	.01016	.01219	.01462	.01754	.02103	.02521	.03021
25	10	.02878	.03448	.04130	.04948	.05927	.07098	.08501
	20	.03691	.04421	.05296	.06343	.07597	.09098	.10894
	30	.04734	.05671	.06792	.08134	.09741	.11664	.13967
60	10	.06825	.08173	.09787	.11720	.14034	.16804	.20119
	20	.08759	.10488	.12559	.15039	.18006	.21559	.25812
	30	.11242	.13461	.16118	.19300	.23107	.27665	.33121
100	10	.11336	.13573	.16252	.19460	.23300	.27896	.33397
	20	.14551	.17422	.20861	.24976	.29903	.35801	.42861
	30	.18679	.22364	.26777	.32059	.38382	.45952	.55012

reason why the MVK two-stage model and extensions of it can fit incidence functions of all human cancers while other models can only fit incidence functions of some human cancers (49). To illustrate how the MVK two-stage model can fit into incidence function of human cancers, we consider the three major incidence functions of human cancers.

Incidence Functions of Female Sex Organs-Breast and Endometrium

For breast cancer of women, the age-specific incidence rates rise steadily until menopause, level off and then continue to rise, albeit more slowly. For this type of incidence curve, the MVK two stage model with logistic growth of normal stem cells and varying $b_2(t) - d_2(t) > 0$ would fit the curve. In fact, Moolgavkar, Day and Stevens (48) have shown that the MVK two-stage model, with appropriate modifications to incorporate the physiologic responses of the breast tissue to menarche and menopause, fits very well the incidence curves of breast cancer in five test populations: Connecticut, Denmark, Finland, Slovenia and Osaka. (See reference 48 for details.) Apparently, the logistic growth pattern of normal stem cells arises from the fact that sex hormones such as estrogen affect the kinetics of growth of nonmalignant breast tissue; the breast grows in response to hormonal stimuli at puberty, and it involutes when these stimuli are removed at menopause. Also, sex hormones act as promoters to affect the cell proliferation rate and differentiation rate through $b_2(t) - d_2(t)$ of initiated cells (46). Thus, the incidence curves of female sex organs are shaped mainly by the functions of sex hormones which on the one hand yield a logistic growth curve for normal stem cells while on the other hand serve as promoters to affect the cell proliferation rate and differentiation rate of initiated cells.

Log-Log Incidence Curves

The incidence rates increase steadily with age so that the log of incidence rate is linearly related to the log of age. Or equivalently, the incidence rate is proportional to the power of age. Examples of human cancers with log-log incidence include many common carcinomas, such as those of the lung, colon, stomach and prostate gland. These were the cancers for which the Armitage–Doll classical multistage model was first proposed. Moolgavkar and Knudson (49) showed that by assuming a Gompertz growth curve for normal stem cells, the MVK homogeneous two-stage model is capable of generating log-log curves.

Gamma-Type Incidence Curves

The incidence rates first increase to achieve a peak sometime in life and then follow by a decline. Examples of these types of incidence curves are provided by retinoblastoma, Wilm's tumor, acute lymphocyte leukemia and Hodgkin's disease. For these cancers, the MVK two-stage model with Gamma density for growth pattern of normal stem cells provides an excellent fit. In these cases, one would normally expect $b_2(t) - d_2(t) < 0$ (49).

3.5 THE PROBABILITY DISTRIBUTION AND MOMENTS OF THE NUMBER OF TUMORS

3.5.1 The Probability Distribution of the Number of Tumors

Let $P_j(t)$ be the probability of having j tumors at time t given N_0 normal stem cells at $t = t_0$. Under conditions given in Theorem (3.2.3), $g(t_0, t) = \psi(1, x_3; t_0, t)$ is obtained from $\psi(t_0, t) = \psi(x_2, x_3; t_0, t)$ in Corollary of Theorem (3.2.3) by setting $x_2 = 1$.

For facilitating computation of $P_j(t)$, it is convenient to write $g(t_0, t)$ as

$$g(t_0, t) = \exp\left\{-q_0(t_0, t) + \sum_{j=1}^{\infty} x_3^j q_j(t_0, t)\right\},$$

where

$$q_j(t_0, t) = (-1)^{\delta_{j0}} N_0 \int_{t_0}^{t} \alpha_1(u) \mu_1(t_0, u) \omega_j(u, t) du,$$

and

$$\omega_j(u, t) = \{\partial^j[\phi(u, t) - 1]/\partial x_3^j\}_{(x_2=1, x_3=0)}.$$

Then, by Theorem (2.1.2), $P_j(t)$ is computed by the following iterative procedure:

$$P_0(t) = \exp[-q_0(t_0, t)] \quad \text{and for} \quad j \geq 1,$$

$$P_j(t) = \sum_{u=0}^{j-1} [(j - u)/j] P_u(t) q_{j-u}(t_0, t). \tag{3.5.1}$$

To compute $P_j(t)$ using (3.5.1), one would need the solution $\phi(t_0,t)$ of equation (3.2.3). As shown in Section (3.3), for the important special cases considered in Section (3.3), the solution $\phi(t_0,t)$ of (3.2.3) is readily available. For these cases, therefore, one may compute $P_j(t)$ for given parameter values. Tan and Gastardo (68) have in fact computed $P_j(t)$ for the special case in which the time interval $[t_0,t]$ is partitioned into nonoverlapping subintervals and in each subinterval one is entertaining a homogeneous MVK two-stage model of carcinogenesis. Given in Figure 3.2 is the probability distribution of the number of tumors for some computer simulated initiation and promotion experiments. The parameter values for these distributions were given in Section (3.4.2)(c) of Section (3.4).

3.5.2 The Cumulants of the Number of Tumors

Given the PGF $g(t_0,t)$ of the number of tumors at time t given N_0 normal stem cells at t_0, one may obtain moments or cumulants of the number of tumors at time t given N_0 normal stem cells at time t_0. In particular, one may obtain the expected value $\mu(t)$ and the variance $V(t)$ of the number of tumors at time t given N_0 normal stem cells at t_0. In this section we

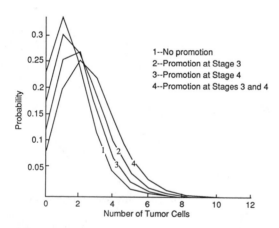

Figure 3.2 The probability density of the number of tumor cells as affected by the regimen of promotion used. Initiation at stage 2 followed by different regimens of promotion with promoter affecting b_2 (parameters as specified in Section *3.4.2)(c)) $C_I = 60$, $C_P = 20$

proceed to obtain $\mu(t)$ and $V(t)$ for the three special cases considered in the previous section.

The Moolgavkar-Venzon-Knudson Two-Stage Model of Carcinogenesis

For this model, $\alpha_i(t) = \alpha_i, i = 1, 2, b_2(t) = b_2$ and $d_2(t) = d_2$; further, one may take $t_0 = 0$. Let $y_2 > y_1$ be defined by $2b_2y_2 = [b_2 + d_2 + \alpha_2 - z\alpha_2] + h(z)$ and $2b_2y_1 = [b_2 + d_2 + \alpha_2 - z\alpha_2] - h(z)$, where $h(z) = [(b_2 + d_2 + \alpha_2 - z\alpha_2)^2 - 4b_2d_2]^{\frac{1}{2}}$.

Define $\Phi(z,t)$ by

$$\Phi(z,t) = \{y_2(1 - y_1) + y_1(y_2 - 1)\exp[(y_2 - y_1)b_2t]\}$$
$$\times \{(1 - y_1) + (y_2 - 1)\exp[(y_2 - y_1)b_2t]\}^{-1}, \tag{3.5.2}$$

Then, from (3.2.7), the PGF $g(z; 0, t) = g(z; t)$ of $T(t)$ given $N(0) = N_0$ is:

$$g(z; t) = \exp\{\alpha_1 \int_0^t N(t - x)[\Phi(z, x) - 1]dx\} \tag{3.5.3}$$

where $N(t - x)$ is the expected number of normal stem cells at time $t - x$ given $N(0) = N_0$.

From (3.5.2), the cumulant generating function (CGF) $C(t)$ of $T(t)$ given $N(0) = N_0$ is:

$$C(t) = \log g[\exp(z), t]$$
$$= \alpha_1 \int_0^t N(t - x)\{\Phi[\exp(z), x] - 1\}dx. \tag{3.5.4}$$

Let $\mu_1(t) = [\partial\Phi(z, t)/\partial z]_{(z=1)}$ and $\mu_2(t) = [\partial^2\Phi(z, t)/\partial z^2]_{(z=1)}$. From (3.5.4), we have:

$$\mu(t) = \alpha_1 \int_0^t N(t - x)\mu_1(x)dx, \tag{3.5.5}$$

and

$$V(t) = \alpha_1 \int_0^t N(t - x)[\mu_1(x) + \mu_2(x)]dx, \tag{3.5.6}$$

From (3.5.2), one may take partial derivatives of $\Phi(z, t)$ with respect to z and set $z = 1$ to obtain $\mu_1(t)$ and $\mu_2(t)$ as

$$\mu_1(t) = (\alpha_2/\epsilon_2)[\exp(t\epsilon_2) - 1], \tag{3.5.7}$$

and

$$\mu_2(t) = 2\exp(t\epsilon_2) \int_0^t \exp(-x\epsilon_2)[b_2\mu_1^2(x) + \alpha_2\mu_1(x)]dx$$

$$= 2(\alpha_2/\epsilon_2)^2\{(b_2/\epsilon_2)\exp(2\epsilon_2 t)$$

$$- [t(b_2 + d_2) - 1]\exp(\epsilon_2 t) - (d_2/\epsilon_2)\}. \tag{3.5.8}$$

An alternate but simpler approach to obtain $\mu_1(t)$ and $\mu_2(t)$ is available by noting from (3.3.3) of Section (3.3.1) that $\Phi(z,t)$ satisfies the following Ricatti equation:

$$\frac{d}{dt}\Phi(z,t) = b_2\Phi^2(z,t) + [z\alpha_2 - (b_2 + d_2 + \alpha_2)]\Phi(z,t) + d_2, \Phi(z,0) = 1. \tag{3.5.9}$$

On both sides of (3.5.9), take partial derivatives with respect to z, set $z = 1$ and note $\Phi(1,t) = 1$. Then, it is immediately seen that $\mu_1(t)$ and $\mu_2(t)$ satisfy the following equations respectively:

$$\frac{d}{dt}\mu_1(t) = \epsilon_2\mu_1(t) + \alpha_2, \mu_1(0) = 0; \tag{3.5.10}$$

and

$$\frac{d}{dt}\mu_2(t) = \epsilon_2\mu_2(t) + 2\mu_1(t)[b_2\mu_1(t) + \alpha_2], \mu_2(0) = 0. \tag{3.5.11}$$

Equations (3.5.10) and (3.5.11) are first order linear equations and can easily be solved to give the solutions as given in (3.5.7) and (3.5.8). Formulas (3.5.5) and (3.5.6) with (3.5.7) and (3.5.8) have been used by Tan and Singh (70) to assess effects of metabolism of carcinogens on cancer tumor development.

The Model with $d_2(t) = 0$

In this case the solution $\phi(t_0,t)$ of (3.2.5) is available from (3.2.16). Let $\Psi(z;t_0,t) = [\xi(z;t_0,t) - \eta(z;t_0,t)]^{-1}$ where $\xi(z;t_0,t) = \exp\{\int_{t_0}^t [b_2(x) + (1-z)\alpha_2(x)]dx\}$ and $\eta(z;t_0,t) = \int_{t_0}^t b_2(y)\xi(z;t_0,y)dy$.

Then, from (3.2.9), the PGF $g(z;t_0,t)$ of $T(t)$ given $N(t_0) = N_0$ is:

$$g(z;t_0,t) = \exp\{\int_{t_0}^t N(x)\alpha_1(x)[\Psi(z;x,t) - 1]dx\}, \tag{3.5.12}$$

where $N(x)$ is the expected number of normal stem cells at time x given $N(t_0) = N_0$. From (3.5.12), one has:

$$\mu(t) = [\partial g(z;t_0,t)/\partial z]_{(z=1)} = \int_{t_0}^{t} N(x)\alpha_1(x)\mu_1(x,t)dx$$

and

$$V(t) = [\partial^2 g(z;t_0,t)/\partial z^2]_{(z=1)} + \mu(t) - [\mu(t)]^2$$
$$= \int_{t_0}^{t} N(x)\alpha_1(x)\mu_2(x,t)dx + \mu(t),$$

where

$$\mu_i(x,t) = [\partial^i \Psi(z;x,t)/\partial z^i]_{(z=1)}, i = 1, 2.$$

To obtain $\mu_i(s,t)$, first observe the following results:

$$\xi(1;s,t) = \exp\{\int_s^t b_2(x)dx\}, \eta(1;s,t) = \int_s^t b_2(y)\xi(1;s,y)dy$$
$$= \int_s^t d\xi(1;s,y) = \xi(1;s,t) - 1$$

by integration by parts;

$$[\partial \xi(z;s,t)/\partial z]_{(z=1)} = -\omega_2(s,t)\xi(1;s,t)$$

and

$$[\partial^2 \xi(z;s,t)/\partial z^2]_{(z=1)} = \omega_2^2(s,t)\xi(a;s,t),$$

where

$$\omega_2(s,t) = \int_s^t \alpha_2(y)dy;$$

noting $\partial \omega_2(s,u)/\partial u = \alpha_2(u)$,

$$[\partial \eta(z;s,t)/\partial z]_{(z=1)} = -\int_s^t b_2(u)\omega_2(s,u)\xi(1;s,u)du$$
$$= -\int_s^t \omega_2(s,u)d\xi(1;s,u)$$
$$= -\omega_2(s,t)\xi(1;s,t) + \int_s^t \alpha_2(u)\xi(1;s,u)du$$

by integration by parts, and

$$[\partial^2 \eta(z;s,t)/\partial z^2]_{(z=1)} = \int_s^t b_2(u)\omega_2^2(s,u)\xi(1;s,u)du$$

$$= \int_s^t \omega_2^2(s,u)d\xi(1;s,u)$$

$$= \omega_2^2(s,t)\xi(1;s,t) - 2\int_s^t \omega_2(s,u)\alpha_2(u)\xi(1;s,u)du$$

by integration by parts.

It follows that

$$\xi(1;s,t) - \eta(1;s,t) = 1, [\partial\xi(z;s,t)/\partial z]_{(z=1)} - [\partial\eta(z;s,t)/\partial z]_{(z=1)}$$

$$= -\int_s^t \alpha_2(u)\xi(1;s,u)du,$$

and $[\partial^2\xi(z;s,t)/\partial z^2]_{(z=1)} - [\partial^2\eta(z;s,t)/\partial z^2]_{(z=1)}$

$$= 2\int_s^t \omega_2(s,u)\alpha_2(u)\xi(1;s,u)du.$$

On substituting these results, one has:

$$\mu_1(x,t) = [\partial\Psi(z;x,t)/\partial z]_{(z=1)} = -\{\partial[\xi(z;x,t) - \eta(z;x,t)]/\partial z\}_{(z=1)}$$

$$= \int_x^t \alpha_2(u)\xi(1;s,u)du$$

and

$$\mu_2(x,t) = [\partial^2\Psi(z;x,t)/\partial z^2]_{(z=1)}$$

$$= 2[\mu_1(x,t)]^2 - 2\int_x^t \omega_2(x,u)\alpha_2(u)\xi(1;x,u)du.$$

The Tan-Gastardo Two-Stage Model

In this model, the time interval $[0,t]$ is partitioned into k nonoverlapping subintervals $L_j = [t_{j-1},t_j), j = 1,\ldots,k-1$ and $L_k = [t_{k-1},t_k]$, with $t_0 = 0$ and $t_k = t$; and in the jth interval, $\alpha_i(t) = \alpha_{ij}, i = 1,2, b_2(t) = b_{2j}$ and $d_2(t) = d_{2j}$.

Let $\epsilon_j = b_{2j} - d_{2j}$ and $\tau_j = t_j - t_{j-1}, j = 1,\ldots,k$ and put for $j = 1,\ldots,k$:

$$f_{1j}(t) = \exp(\epsilon_j t) \text{ with } A_j = f_{1j}(\tau_j),$$

$$f_{2j}(t) = (\alpha_{2j}/\epsilon_j)[exp(\epsilon_j t) - 1] \text{ with } B_j = f_{2j}(\tau_j),$$

and

$$\mu_{1j}(t) = f_{1j}(t) \sum_{u=j+1}^{k} B_u[\prod_{v=j+1}^{u-1} A_v] + f_{2j}(t)$$

with $\mu_{1j}(\tau_j) = W_j$, where $W_{k+1} = 0$, $\prod_{v=1}^{0}$ is defined as 1 and $\sum_{u=k+1}^{k}$ is defined as zero.

Then, as shown in Appendix (3A),

$$\mu(t) = \sum_{j=1}^{k} \alpha_{1j} \int_{0}^{\tau_j} N_j(x)\mu_{1j}(\tau_j - x)dx, \qquad (3.5.13)$$

where $N_j(x)$ is the expected number of normal stem cells at $t_j + x$ given $N(0) = N_0$.

Note that with $W_{k+1} = 0$ and $W_j = \mu_{1j}(\tau_j)$, one has for $j = 1,\ldots,k$:

$$W_j = \mu_{1j}(\tau_j) = f_{1j}(\tau_j) \sum_{u=j+1}^{k} B_u[\prod_{v=j+1}^{u-1} A_v] + f_{2j}(\tau_j)$$

$$= \sum_{u=j+1}^{k} B_u[\prod_{v=j}^{u-1} A_v] + B_j = \sum_{u=j}^{k} B_u[\prod_{v=j}^{u-1} A_v].$$

To obtain $V(t)$, put, for $j = 1,\ldots,k$:

$h_{1j}(t) = (2b_{2j}/\epsilon_j) \exp(\epsilon_j t)[\exp(\epsilon_j t) - 1]$ with $h_{1j}(\tau_j) = H_j$,

$h_{2j}(t) = (2\alpha_{2j}/\epsilon_j^2) \exp(\epsilon_j t)[-(b_{2j} + d_{2j})\epsilon_j t - 2b_{2j} + 2b_{2j} \exp(\epsilon_j t)]$

with $h_{2j}(\tau_j) = G_j$,

$h_{3j}(t) = (\alpha_{2j}^2/\epsilon_j^3)\{\epsilon_j[\exp(\epsilon_j t) - 1] - 2(b_{2j} + d_{2j})$

$\times[\epsilon_j t \exp(\epsilon_j t)] + 2(b_{2j} + d_{2j})[\exp(\epsilon_j t) - 1]^{-1}\}$ with $h_{3j}(\tau_j) = F_j$;

further let $U_{k+1} = 0$ and for $j = 1,\ldots,k$,

$$U_j = \sum_{u=j}^{k-1} (H_u W_{u+1}^2 + G_u W_{u+1} + F_u) \prod_{v=j}^{u-1} A_v + F_k[\prod_{v=j}^{k-1} A_v],$$

where $\sum_{u=k}^{k-1}$ is defined as 0 and $\prod_{v=k}^{k-1}$ defined as 1. Then, as shown in Appendix (3A),

$$V(t) = \sum_{j=1}^{k} \alpha_{1j} \int_{0}^{\tau_j} N_j(x)\mu_{2j}(\tau_j - x)dx + \mu(t), \qquad (3.5.14)$$

where

$$\mu_{2j}(t) = h_{1j}(t)W_{j+1}^2 + h_{2j}(t)W_{j+1} + h_{3j}(t) + f_{1j}(t)U_{j+1}.$$

Note that $\mu_{2j}(\tau_j) = U_j, j = 1, \ldots, k$ so that U_j is obtained by the following iterative relation

$$U_j = H_j W_{j+1}^2 + G_j W_{j+1} + F_j + A_j U_{j+1} \text{ with } U_{k+1} = 0.$$

Formula (3.5.13) and (3.4.14) were used by Tan and Singh (72) to assess effects of metabolism of carcinogens on cancer tumor development for the Tan-Gastardo model.

3.6 INTERPRETATION OF BIOLOGICAL PHENOMENA BY THE TWO-STAGE MODELS OF CARCINOGENESIS

The usefulness of mathematical models is to provide a basis to interpret fundamental biological phenomena and to suggest experiments. In this section I shall proceed to demonstrate that the MVK two-stage model and extensions of it provide satisfactory explanations for most of the biological phenomena of carcinogenesis. To illustrate, in this section I shall proceed to consider some outstanding issues of carcinogenesis: The initiation-promotion phenomenon, the hormone effects, the familial human cancers, and the intermediate lesions. In what follows, I shall show that the MVK two-stage model and its extensions provide satisfactory explanations of these biological phenomena.

The Initiation and Promotion Phenomenon

As summarized in Chapter One, there are several important features concerning initiators and promoters. Specifically, one may note the phenomenological events: (i) Initiators are mutagens, (ii) Promoters are effective in inducing tumors only if applied after the tissues have been exposed to initiators, (iii) Initiation is irreversible and heritable but promotion is reversible, (iv) Promoters are in general non-mutagens. In the framework of the two-stage models, the fundamental characteristics of initiators and promoters imply that initiators affect the rate of the first event while promoters facilitate the increase of the proliferation rate of intermediate cells by either increasing $b_2(t)$ or decreasing $d_2(t)$ or both.

Thus, the effects of initiators are to increase the probability of the first mutational type of event leading to mutations (in the broad sense) from normal stem cells to intermediate cells. On the other hand, promoters act to increase $b_2(t) - d_2(t)$ resulting in an increased number of intermediate cells. The following experiments provide strong support for the above formulation:

i.   Zarbl, Sukumar and Barbacid (80) reported that, by injecting nitrosomethylurea (NMU) into the breast of female rats, NMU binds with DNA, inducing a G(guanine) to A(adenine) base transition at the second nucleotide of codon 12 of the ras oncogene. This point mutation leads to initiation of carcinogenesis.

ii. Many initiation-promotion experiments (see reference 79) demonstrated that mouse skin, when first treated by an initiator such as 7, 12-dimethylbenz[a]anthracene (DMBA) and followed by a promoter such as 12-0-tetradecanoylprobol-13-acetate (TPA), gives rise to papillomas. Papilloma is an intermediate premalignant lesion, which may either regress or progress with a very low rate of conversion to yield squamous cell carcinomas (malignant conversion); however, Hennings et al. (24) reported that initiators such a N-methyl-N'-nitroso-guanidine (MNNG) or 4-nitroquinoline-N-oxide (r-Q0), but not promoters, would induce carcinomas from papillomas. Similar initiation-promotion-initiation experiments have also been reported by Scherer et al. (62) on rat liver which was in fact predicted by Potter in 1980 (60). These results imply that carcinogenesis starts with a mutational type of event which is initiated by initiators in initiation-promotion experiments; cells with this genetic change would then expand clonally to develop intermediate lesions (papillomas in the case of rat skin) by the action of promoters. A further genetic change is required for the irreversible progression of the intermediate lesion (papillomas in the case of rat skin) to malignant tumors(carcinomas).

In the a bove formulation, it is then apparent that promoters are effective only after treatment by an initiator; for otherwise promoters would have no target cells to act on. Also, as demonstrated by Tan and Gastardo (68), prolonged application of an initiator would facilitate both critical events and would lead eventually to malignant transformation. In fact, Iversen (27) has shown that after prolonged exposure, pure initiator urethane alone is capable of inducing carcinomas in mouse skin.

Effects of Hormones on Carcinogenesis

Moolgavkar (46) has demonstrated that the two-stage models could be used to relate actions of sex hormones with the risk factors of breast and endometrium cancers of women. It appears that some sex hormones affect breast and endometrium cancers basically by two avenues: Stimulation of growth of normal stem cells and as promoters for cell proliferation and differentiation of initiated cells. The first action increases the pool of susceptible normal stem cells so that the chances of initiation by some carcinogens are increased. The second event expands clonally the intermediate cells which have sustained the first mutation. The following features provide some basic guidelines which would explain some specific characteristics of breast and endometrium cancers. To illustrate, consider specifically the breast cancer of women.

i. Some sex hormones such as estrogen stimulate growth of normal stem cells. Since the level of estrogen is high around puberty and involutes after menopause, the growth of breast tissue rises sharply around puberty, holds steady during most of the life time and then slows down after menopause. This yields a logistic growth curve for the breast of women.

ii. Estrogen acts to increase cancer risk by increasing susceptibility to cancer initiation and by promoting cell proliferation of normal stem cells and initiated cells (4,46). On the other hand, progesterone acts to decrease cancer risk by inactivating carcinogens and by increasing cell differentiation of normal stem cells and initiated cells (4,46). Or, as promoters, estrogen would increase both $b_1(t)$ and $b_2(t)$ while progesterone would incresase both $d_1(t)$ and $d_2(t)$; the overall effect of estrogen as a promoter is therefore to increase $b_1(t) - d_1(t)$ and $b_2(t) - d_2(t)$ while the overall effect of progesterone as a promoter is to decrease $b_1(t) - d_1(t)$ and $b_2(t) - d_2(t)$, yielding a protective effect against cancer.

Taking into account the above features, Moolgavkar (46) showed that the two-stage models provide a logical explanation for many risk factors. To illustrate, consider two important factors: Early menarche and late menopause, and early full-term pregnancy.

i. It is well-known that early menarche and late menopause increase the risk of breast cancer of women. Under the premise that menses increase and maintain estrogen levels, the results are well explained by the two-stage models. Observe that early menarche leads to an early growth of the breast and to increase the number of normal stem cells early in life while late menopause delays involution of the breast. Thus, the effect of both early menarche and late menopause is to increase the time period during which a large number of normal stem cells is present. Moolgavkar, Day, and Stevens (48) have shown that the MVK two-stage model predicts that menarche at age 10 carries a two-fold risk over menarche at age 16, and that menopause at age 52 carries a two-fold risk over menopause at age 42. These predictions appear to be in good agreement with the observations.

ii. It is well known that early full-term pregnancy confers a protective effect against breast cancer of women. Under the premise that early full-term pregnancy maintains a high level of progesterone, the phenomenon is well explained by the two-stage models. Since progesterone promotes cell differentiation, during early full-term pregnancy, a fraction of normal stem cells and intermediate cells have undergone terminal differentiation so that the pool of susceptible normal stem cells and the pool of intermediate cells have decreased. It follows that the mutation rates of the first and second events in two-stage models are also reduced. Taking this hypothesis into account, Moolgavkar (46) showed that the two-stage models provided a good quantitative description of the data from a large international study of MacMahon et al. (39).

Familial Human Cancers

For many human cancers, individuals whose relatives have cancer are usually at higher risk. These human cancers have been referred to as familial human cancers. Examples of familial human cancers include, among many others, breast cancer of women, colon cancer, retinoblastoma and Wilm's tumor. In terms of the two-stage models, the phenomenon of clustering of cancer cases in families can be explained either through genetic factors or through epigenetic factors. Epigenetically, the clustering of cancer cases in families is due to the fact that members in the same family share, to some extent, similar risk factors for cancers. Genetically, cancer genes may pass on from parents to children so that there is a higher probability for relatives to share the same cancer gene than for nonrelatives. To illustrate how the genetic factors may relate to familial human cancers, consider the following specific examples:

i. For retinoblastoma, it is well known that there is a Rb gene at chromosome 13q14. Let rb (recessive) be the mutated form of Rb gene.

Then individuals who are homozygotes or hemizygotes for Rb develop the cancer phenotype. Now, for this cancer, the first event may occur either in germline cells in which case it is hereditary, or in somatic cells in which case it is nonhereditary (12,13,33). If the first event (mutation from wild allele Rb to rb) occurs in germline cells, then all stem cells at birth are initiated cells. In this case, only one stage is required for the development of cancer.

Suppose now a mutation has occurred in a germline cell. Then one of the parents is a rb gene carrier (heterozygote). There is a chance that this rb gene will pass on from parent to children and the probability that two children in the same family carry the rb gene is 1/4. For these two children only one stage is required for the development of cancer. If the mutation rate for the first event is 10^{-7}, then the risk for these two children is 10^7 times those of nonrelatives who do not carry the rb gene.

The above demonstrates how retinoblastoma clusters in family members (see 12, 13, 18). A similar explanation for family clusters of cancer cases applies to Wilm's tumor and colon cancer, among many others. For Wilm's tumor it is known that there is a Wm gene in chromosome 11p13 (22, 34, 54) while for colon cancer an antioncogene Fap has recently been identified in chromosome 5q21 (7,65)

ii. It is known that there are accessory cancer genes which affect carcinogenesis indirectly by either affecting the mutation rates of the first and/or second event or by affecting cell proliferation and differentiation rates of initiated cells, or both. Examples of accessory cancer genes include, among many others, xeroderma pigmentosum (xp), ataxia telangiectasia (at), Franconi's anemia (fa) and Bloom's syndrome. To illustrate how the accessory cancer genes are related to the clustering of cancer cases in families, consider the inherited disease xeroderma pigmentosum. It has been reported that for patients with the xp gene, DNA repair for damages by ultraviolet light is defective (64); the afflicted persons are therefore remarkably sensitive to sunlight, the most lethal effect being skin cancers. Within the context of the two-stage model, the xp gene increases the mutation rates of both events in the presence of UV light. In the case of accessory genes, it is therefore anticipated that cancer cases cluster in families. For example, the probability that relatives of xp patients will carry the xp gene is obviously high as compared with nonrelatives.

Intermediate Lesions

In the framework of the two-stage models, it is anticipated that clonal expansion of intermediate cells would lead to intermediate lesions. As shown by Moolgavkar and Knudson (49), intermediate lesions have in fact been observed in many cancers. For example, in chemical carcinogenesis, papillomas of mouse skin and enzyme-altered loci in the rat liver have been identified as intermediate lesions; in human beings, polyps of the colon and locules in the liver (20,21) were identified as intermediate lesions. Also, C-cell hyperplasia of the thyroid gland is likely to be an intermediate lesion in the carcinoma of the thyroid gland and lobular carcinoma in situ a possible intermediate lesion in breast cancer of women.

Based on the two-stage models, one expects that the intermediate lesions are clonal and sporadic in nonhereditary cases and polyclonal in hereditary cases. Many observed results appear to agree with this projection (see reference 49). For example, Hsu et al. (26) have shown that polyps in Gardner's syndrome, a dominantly inherited polyposis syndrome, are polyclonal.

3.7 RISK ASSESSMENT BY THE TWO-STAGE MODELS OF CARCINOGENESIS

In the past, the major model for assessing the risk of environmental agents is the classical Armitage–Doll multistage model of carcinogenesis. As we have demonstrated in Chapter One, however, biologically this model is not plausible. This has caused confusion in risk assessment and is a subject of considerable dispute (6). To quote Albert (1), "Linear low dose extrapolation, using the linearized multistage model, is currently the dominant approach to assessing cancer risks in Federal regulatory agencies; this is true in spite of the fact that its use is recommended by the Office of Science and Technology Policy (1985) "Cancer Principles" and the EPA Guidelines for Carcinogen Risk Assessment (1985) only when there are no other acceptable alternative extrapolation models. The risk estimates obtained by the use of the linearized multistage model are described as plausible upper limit estimates in the sense that the true risk are not likely to be greater but could be considerable smaller, approaching zero. The great uncertainty in the risk estimates obtained by the linearized multistage model make them very difficult to use in the regulatory management of carcinogens."

Thorslund, Brown and Charnley (74) have also shown that the classical Armitage–Doll multistage model give inconsistent and unstable results. Because of these and the unrealistic nature of the Armitage–Doll model, they proposed using the Moolgavkar-Vernzon-Knudson (MVK) two-stage model to assess risks of environmental agents. The applications of the two-stage models to risk assessments of environmental agents have also been suggested by Moolgavkar, Dewanji, and Venzon (50), Portier and Bailer (58) and Tan and Starr (73). This opens up a new research frontier in risk assessment of environmental agents.

3.7.1 Risk Assessment by the Quantal Response Method

In this approach, similar animals (rats for example) are exposed to carcinogens with different levels of doses and observations on the number of tumors or deaths are made after some fixed time. For such data typical procedures for risk assessment are given as follows:

i. Assume a model (two-stage model, for example) and develop a dose-response curve $P(d)$, i.e. the probability of response (tumor in cancer cases) as a function of dose d of carcinogen.

ii. Estimate parameters by the maximum likelihood method by maximizing the log likelihood

$$L \propto \sum \{x_i \log P(d_i) + (n_i - x_i) \log[1 - P(d_i)]\},$$

where x_i is the number of responses among n_i animals after a fixed time when treated with dose d_i.

iii. Substituting the estimated parameter values into $P(d)$ to obtain $\hat{P}(d)$ and obtain VSD (Virtually Safe Dose) d_0 from $\hat{P}(d_0) = \epsilon$ for some given small ϵ. (This procedure is often referred to as low-dose extrapolation.)

iv. Convert d_0 into an equivalent dose for human beings (Species Extrapolation).

Assuming a one-hit model, multi-hit model, Weibull model and Armitage–Doll model and using the above procedures, Van Ryzin (76,77) showed that different models gave very different results in risk assessment. It follows that in risk assessment of environmental agents, the assumed model should be as close as possible to the true biological mechanisms. This suggests that the MVK two-stage model and extensions of it may be used to assess the risk of environmental agents since biologically it is a reasonable model to use.

In the application of the MVK two-stage model and extensions of it to assess risk of environmental agents, note that for fixed t and dose d_i,

$$P(d_i) = P(d_i,t) = 1 - exp\{- \int_{t_0}^{t} \lambda(d_i,x)dx\},$$

where $\lambda(d_i,t)$ is the incidence function at t of cancer tumors for given dose d_i. Under different assumptions of the actions of carcinogens, $\lambda(d_i,t)$ can be obtained by results in Section (3.4) for the important cases. Research in this direction is at the beginning stage, however.

3.7.2 Risk Assessment by the Time to Events Approach

In this approach, similar animals (rats or mice, for example) are exposed to an environmental agent in question with different dose (d) levels ($d = 0$ corresponds to control). Then, times to death are recorded and autopsies made to determine if tumors exist. Simultaneously, random or scheduled sacrifices are made at different times. Given these animal experiments with sacrifices, at each death time (natural or sacrifice), there are four types of observations: The number of natural deaths with tumor, the number of natural deaths without tumor, the number of sacrifices with tumor and the number of sacrifices without tumor. For analyzing these types of data, recently nonparametric methods (14,15,43), semiparametric methods (16,56,57) and parametric methods (17,59) were developed. In the non-parametric approach, estimates of survival probabilities for both cancer tumors and competing risk of other causes of death are basically based on Kaplan-Meier type of estimates (14,15,43). In the semi-parametric approach, while the competing risk is based on Kaplan-Meier type of estimates, the Weibull incidence (Armitage–Doll multistage model) was used for carcinogenesis (16,56,57). In the parametric approaches, while the Weibull incidence was adopted for modeling carcinogenesis, Weibull or logistic models have been used for competing risks of other causes of death (17,59). Intuitively, if the models are correct, the semi-parametric approach should be more efficient than the nonparametric approach while the parametric approach should be more efficient than the semi-parametric approach; a formal proof of this statement is available from Efron (19) for logistic models. In Chapter 1, it is shown that the Armitage–Doll model is not plausible from a biological viewpoint; it has also been demonstrated that the Armitage–Doll model gave inconsistent and unreliable results in risk assessment. It follows that in the semi-parametric and parametric approaches, it is imperative to in-corporate biologically supported carcinogenesis models to assess risks of environmental agents. Recently attempts have been made by Portier and Bailer (58), Moolgavkar and Luebeck (51), and Tan and Starr (73) to incorporate the two-stage models into risk assessment fo environmental agents. Research in this direction is just beginning, however.

3.7.3 Classifications of Carcinogens by Using Two-Stage Models of Carcinogenesis

To develop strategies for cancer prevention and control, it is desirable to classify carcinogens according to their action. As suggested by reports of Krailo, Thomas and Pike (35), this could be achieved by applying the MVK two-stage model and extensions of it. Note that there are at least six types of carcinogens: cocarcinogens, initiators, promoters, complete carcinogens, completers and inhibitors. In practice, very often a carcinogen may be classified into more than one category. We now elaborate on these classifications.

Cocarcinogens

Cocarcinogens are carcinogens which increase the cell proliferation rate $b_1(t) - d_1(t)$ of normal stem cells, thus increasing the number of target normal stem cells for the initiation process of carcinogenesis. An example of these carcinogens is the dietary sodium chloride for gastric cancer. In an effort to replace the cells necrotized by sodium chloride, normal stem cell proliferation within the gastric mucosa increases, thus raising the number of actively dividing cells that are susceptible to initiation and increasing the probability that an initiated cell will be formed. As another example, estrogen, besides being a promoter for initiated cells, stimulates the growth of the breast tissue of women, thus increasing the number of targeted normal stem cells for the initiation process of carcinogenesis.

Initiators

Carcinogens which act to increase the rate of the first event are called initiators. The first event may be either point mutation, or oncogene activation, or chromosome translocation. Examples of initiators include MNNG, DMBA, NMU and Aflatoxin B1 among others.

Promoters

Carcinogens which act to increase the rate of cell proliferation of initiated cells are called promoters. Examples of promoters include TPA, croton oil, benzo(e)pyrene, benzol peroxide and tobacco smoke among others.

Complete Carcinogens

Complete carcinogens are those carcinogens which are both an initiator and a promoter. Examples of complete carcinogens include benzo(e)pyrene and tobacco smoke among others.

Completers

Carcinogens which act mainly to increase the rate of the second event are called completers. Experiments on rat skin by Hennings et al. (24) and on rat liver by Sheer et al. (62) indicate that many initiators are also completers. In the inherited disease Bloom's syndrome, the accessory gene increases the frequency of sister chromatid and homologous chromosome exchanges, thus predisposing the patient to leukemia and other cancers. In this case, the accessory gene in Bloom's syndrome acts as a completer since homologous chromosome exchanges could lead to homozygosity at an antioncogene locus in a cell that has suffered the first hit (45,49).

Inhibitors

Agents which act to reduce cancer rates are called inhibitors. According to this definition, may antioxidants such at butylated hydroxy-toluene (BHT) and butylated hydroxyanisode (BHA) are inhibitors. Nontoxic compounds, such as retinoids (37) and some vitamins such as Vitamin E and Vitamin D (61,63) are inhibitors since these compounds have been shown to induce cell differentiation in metaplastic tissue.

Using the MVK two-stage model and extensions of it, one may proceed to develop statistical procedures to classify carcinogens according to their action. While statistical procedures remain to be developed, some preliminary work by Krailo, Thomas and Pike (35) with some case-control data of breast cancer of women in Los Angeles suggest that by just comparing likelihood values, it is possible to classify carcinogens. Note that this type of research is extremely important for cancer prevention and control since the basic procedures of prevention and control depend on actions of carcinogens. This type of research is at the beginning stage, however.

3.8 CONCLUSIONS AND SUMMARY

This chapter is devoted to the mathematical theories of the MVK two-stage model and the extensions of it and applications of these models. These models specify that cancer tumors develop from normal stem cells after two consecutive heritable genetic changes and that the normal stem cells and the intermediate cells are subjected to stochastic cell proliferation and cell differentiation. An important feature of these models is that the first event occurs either in germline cells or in somatic cells while the second event always occurs in somatic cells.

As illustrated in Section (2.1), for cancers related to antioncogenes, the two-stage models provide a mathematical description of the biological mechanism. For cancers related to oncogenes, the two-stage models are also appropriate if the combination of immortalization and transformation is sufficient to induce tumorigenic conversion of normal stem cells.

For developing mathematical theories of the two-stage models, in Section (3.2) we give some procedures for computing the probability generating functions (PGF) of the numbers of intermediate cells and tumor cells under some general conditions. The PGF of the number of tumor cells under some general conditons is given in formula (3.2.7) which shows that under some general conditions the two- stage models of carcinogenesis are in fact filtered Poisson processes as defined in Chapter Two. By using formula (3.2.7), in Section (3.4) we then proceed to derive the age-specific incidence functions of tumors and provide an approximation to these incidence functions under general conditions. Using formula (3.2.7), in Section (3.5) we also develop a procedure for computing the probabilities of the number of tumors and derive formulas for the means and the variances of the number of tumors. Throughout this chapter, we illustrate the theories by three important specifc two-stage models - the MVK two-stage model, the model with no death and the Tan-Gastardo model. The MVK model is a time-homogeneous model in which the mutation rates and the birth rate and the death rate of intermediate cells are assumed to be constants while the growth of the normal stem cells are deterministic functions. The Tan-Gastardo model is an extension of the MVK two-stage model, in which the time interval is partitioned into nonoverlapping intervals in each of which one assumes a MVK two-stage model.

To illustrate the usefulness of the two-stage models, we demonstrate in Section (3.6) how to use the two-state models to interpret some important biological phenomena which include the initiation-promotion phenomenon, the hormone effects in breast cancer of women, the familial human cancers, and the intermediate lesions. It appears that the initiation-promotion phenomenon is well explained by the two-stage models with the specification that initiators affect the rate of the first event while promoters facilitate the increase of the proliferation rate of intermediate cells. The specific characteristics of breast and endometrium cancers of women are well explained by the two-stage models because estrogen (female hormone) acts to increase susceptibility to cancer initiation and to promote cell proliferation of normal stem cells and intermediate cells while progesterone acts to inactivate carcinogens and to promote cell differentiation of normal stem cells and intermediate cells. The familial human cancers are explained by the two-stage models either by the feature that the first event of the two-stage models may occur in germline cells or by the fact that members in the same family share similar risk factors for cancers. Finally in Section (3.7), we demonstrate that the two stage models may be used to assess risks of environmental agents and to classify carcinogens according to its application; research in this area is just beginning, however.

APPENDIX

Let $\psi(t_k) = \psi(u, v; t_k)$ be the PGF of intermediate cells and tumor cells at time t_k given $N(0) = N_o$ and let $\phi_j(\tau_j) = \phi_j(u, v; \tau_j)$ be the PGF of intermediate cells and tumors at time $t = t_j$ given one intermediate cell at time t_{j-1}. Let $y_{2j} > y_{1j}$ be defined by:

$$2b_{2j}y_{2j} = [b_{2j} + d_{2j} + \alpha_{2j} - v\alpha_{2j}] + R_j(v)$$

and $2b_{2j}y_{1j} = [b_{2j} + d_{2j} + \alpha_{2j} - v\alpha_{2j}] - R_j(v),$

where $R_j(v) = [(b_{2j} + d_{2j} + \alpha_{2j} - v\alpha_{2j})^2 - 4b_{2j}d_{2j}]^{\frac{1}{2}}$.

Then, as in Section (3.2),

$$\phi_j(u, v; \tau_j) = \phi_j(\tau_j) = \{y_{2j}(u - y_{1j}) + y_{1j}(y_{2j} - u)exp[(y_{2j} - y_{1j})b_{2j}\tau_j]\}$$

$$\times \{(u - y_{1j}) + (y_{2j} - u)exp[(y_{2j} - y_{1j})b_{2j}\tau_j]\}^{-1}. \tag{A.3.1}$$

Let $g_k(x) = \phi_k(u,v;x)$ and for $j = 1,2,\ldots,k-1$, define $g_j(x) = \phi_j[g_{j+1}(\tau_{j+1}),v;x]$. Then, given in formula (3.3.11) of Section (3.3), we have, to order $O(N_o^{-1})$:

$$\psi(t_k) \cong exp\left\{\sum_{j=1}^{k}\alpha_{1j}\int_o^{\tau_j}N_j(x)[g_j(\tau_j-x)-1]dx\right\}. \tag{A.3.2}$$

By taking the derivative with respect to v over $\psi(t_k)$ and putting $u = v = 1$, one has:

$$\mu(t_k) = \sum_{j=1}^{k}\alpha_{ij}\int_o^{\tau_j}N_j(x)\mu_{1j}(\tau_j-x)dx$$

where $\mu_{1j}(t) = [\partial g_j(t)/\partial v]_{(u=v=1)}$. Now, $g_j(t) = 1$ if $u = v = 1$ for all $j = 1,\ldots,k$. By the chain rule,

$$[\partial g_j(t)/\partial v]_{(u=v=1)} = \{\partial\phi_j[g_{j+1}(\tau_{j+1}),v;t]/\partial v\}_{(u=v=1)}$$
$$= [\partial\phi_j(u,1;t)/\partial u]_{(u=1)}[\partial g_{j+1}(\tau_{j+1})/\partial v]_{(u=v=1)} + [\partial\phi_j(1,v,t)/\partial v]_{(v=1)}.$$

From (A.3.1), $f_{1j}(t) = [\partial\phi_j(u,1;t)/\partial u]_{u=1} = \exp(\epsilon_j t)$ and

$$f_{2j}(t) = [\partial\phi_j(1,v;t)/\partial v]_{(v=1)} = (\alpha_{2j}/\epsilon_j)[\exp(\epsilon_{jt})-1].$$

Hence, with $\mu_{1j}(\tau_j) = W_j$ and $W_{k+1} = 0$, $\mu_{1j}(t) = f_{1j}(t)W_{j+1}+f_{2j}(t)$, and $W_j = \mu_{1j}(\tau_j) = f_{1j}(\tau_j)W_{j+1}+f_{2j}(\tau_j) = A_jW_{j+1}+B_j$, where $A_j = f_{1j}(\tau_j)$ and $B_j = f_{2j}(\tau_j)$.

By mathematical induction, the relationship $W_j = A_jW_{j+1}+B_j$ with $W_{k+1} = 0$ leads to

$$W_j = \sum_{u=j}^{k}B_u[\Pi_{v=j}^{u-1}A_v], j = 1,\ldots,k-1.$$

To obtain $V(t)$, observe first that

$$V(t) = [\partial^2\psi(t)/\partial v^2]_{(v=v=1)} + \mu(t) - \mu^2(t)$$
$$= \sum_{j=1}^{k}\alpha_{1j}\int_o^{\tau_j}N_j(x)\mu_{2j}(\tau_j-x)dx + \mu(t),$$

where $\mu_{2j}(t) = [\partial^2 g_j(t)/\partial v^2]_{(u=v=1)}$. Noting that $g_j(t) = 1$ if $u = v = 1$ for all $j = 1,\ldots,k$, by the chain rule, with $\mu_{1j}(t) = \mu_{1j}(t)$,

$$\mu_{2j}(t) = [\partial^2 g_j(t)/\partial v^2]_{(u=v=1)}$$

$$= [\partial^2 \phi_j(u,1;t)/\partial u^2]_{(u=1)} \mu_{1j+1}^2(\tau_j+1)$$
$$+ 2[\partial^2 \phi_j(u,v;t)/\partial u \partial v]_{(u=v=1)} \mu_{1j+1}(\tau_j+1)$$
$$+ [\partial \phi_j(u,1;t)/\partial u]_{(u=1)} [\partial^2 g_{j+1}(\tau_j+1)/\partial v^2]_{(u=v=1)}$$
$$+ [\partial^2 \phi_j(1,v;t)/\partial v^2]_{(v=1)}$$
$$= h_{1j}(t) W_{j+1}^2 + h_{2j}(t) W_{j+1} + f_{1j}(t) U_{j+1} + h_{3j}(t),$$

where

$$h_{1j}(t) = [\partial^2 \phi_j(u,1;t)/\partial u^2]_{(u=1)},$$
$$h_{2j}(t) = 2[\partial^2 \phi_j(u,v;t)/\partial u \partial v]_{(u=v=1)},$$
$$h_{3j}(t) = [\partial^2 \phi_j(1,v;t)/\partial v^2]_{(u=1)}$$

and $U_j = \mu_{2j}(\tau_j)$.

Obviously $U_{k+1} = 0$ and from $\phi_j(u,v,t)$ given in (A.3.1), one obtains $h_{1j}(t), h_{2j}(t)$ and $h_{3j}(t)$ as given in Section (3.5).

Put $h_{1j}(\tau_j) = H_j, h_{2j}(\tau_j) = G_j, h_{3j}(\tau_j) = F_j$ and note $f_{1j}(\tau_j) = A_j$. Then with $U_{k+1} = 0, U_j = \mu_{2j}(\tau_j) = h_{1j}(\tau_j) W_{j+1}^2 + h_{2j}(\tau_j) W_{j+1} + f_{1j}(\tau_j) U_{j+1} + h_{3j}(\tau_j) = H_j W_{j+1}^2 + G_j W_{j+1} + A_j U_{j+1} + F_j, j = 1,2,\ldots,k.$

Noting this, we have by mathematical induction:

$$U_j = \sum_{u=j}^{k-1} (H_u W_{u+1}^2 + G_u W_{u+1} + F_u)[\Pi_{v=j}^{u-1} A_v] + F_k [\Pi_{u=j}^{k-1} A_v], j = 1,\ldots,k.$$

REFERENCES

1. Albert, R.A., The time to tumor approach in risk assessment, In: Mechanism of DNA Damage and Repair, Implications for Carcinogenesis and Risk Assessment; Eds. M.G. Simic, L. Grossman and A.C. Upton,, Plenum Press, New York, 1986.

2. Armitage, P. and Doll, R., A two-stage theory of carcinogenesis in relation to the age distribution of human cancer, Brit. Jour. Cancer **11** (1957), 161-169.

3. Armitage, P. and Doll, R., Stochastic models for carcinogenesis, In: Fourth Berkeley Symposium on Mathematical Statistics and Probability., 19-38, Univ. California Press, Berkeley, CA, 1961.

4. Armstrong, B., Endocrine factors in human carcinogenesis, IARC Scientific Publication, No. 39, Lyon, France, 1982.

5. Barrett, J.C. and Fletcher, W.F., Cellular and molecular mechanisms of multistep carcinogenesis in cell culture models. In: "Mechanisms of Environmental Carcinogenesis. Vol, II. Multistep models of carcinogenesis", edited by Barrett, J.C., CRC Press, Boca Raton, 1986.

6. Bekkum, Van, D.W. and Bentvelzen, P., The concept of gene transfer-misrepair mechanism of radiation carcinogenesis may challenge the linear extrapolation model of risk estimation for low radiation doses, Health Phys. **43** (1982), 231-237.

7. Bodmer, W.F. et al., Localization of the gene for familial adenomatous polyposis on chromosome 5, Nature **328** (1987), 614-616.

8. Buckley, I., Oncogenes and the nature of Malignancy, Adv. in Cancer Res. **50** (1988), 71-93.

9. Buick, R.N. and Pollak, M.N., Perspective on clonogenic tumor cells, stem cells and oncogenes, Cancer Res. **44** (1984), 4909-4918.

10. Cairns, J. and Logan, J., Step by step into carcinogenesis, Nature **304** (1983), 582-583.

11. Carter, R.L. (ed.), Precancerous States, Oxford Univ. Press, London, England, 1984.

12. Cavenee, W.K. et al., Expression of recessive alleles by chromosomal mechanisms in retinoblastoma, Nature **305** (1983), 719-784.

13. Cavenee, W.E. et al., Genetic origin of mutations predisposing to retinoblastoma, Science **228** (1985), 501-503.

14. Dewanji, A. and Kabfleish, J., Nonparametric methods for survival/sacrifice experiments, Biometrics **42** (1986), 325-342.

15. Dinse, D.E, Nonparametric prevalence and mortality estimators for animal experiments with incomplete cause- of-death data, Jour. Amer. Statist. Assoc. **81** (1986), 328-336.

16. Dinse, D.E., Estimating tumor incidence rates in animal carcinogenicity experiments, Biometrics **44** (1988), 405-416.

17. Dince, D.E., Simple parametric analysis of animal tumorigenicity data, Jour. Amer. Statist. Assoc. **83** (1988), 238-649.

18. Dryja, T.P. et al., Homozygosity of chromosome 13 in retinoblastoma, The New England J. Medicine **310** (1984), 550-553.

19. Effron, B., Logistic regression, survival analysis and the Kaplan-Meier curve, Jour. Amer. Statist. Assoc. **83** (1988), 414-425.

20. Farber, E., Cellular biochemistry of the stepwise development of cancer with chemicals: G.H.A Clowes Memorial Lecture., Cancer Research **44** (1984), 5463-5474.

21. Farber, E., Experimental induction of hepatocellular carcinoma as a paradigm for carcinogenesis, Clin. Physical. Biochem. 5 (1987), 152-159.

22. Fearson, E.R., Volgestein, B. and Feinberg, A.P., Somatic deletion and duplication of genes on chromosome 11 in Wilm's tumors, Nature **309** (1984), 174-176.

23. Fung, J.K. et al, Structural evidence for the authenticity of the human retinoblastoma gene, Science **236** (1987), 1657-1660.

24. Hennings, H. et al., Malignant conversion of mouse skin tumours is increased by tumor initiators and unaffected by tumour promoters, Nature **304** (1983), 67-69.

25. Horowitz, J. et al., Point mutational inactivation of the retinoblastoma antioncogene, Science **243** (1989), 937-940.

26. Hsu, S.H. et al., Multiclonal origin of polyps in Gardner syndrome, Science **221** (1983), 951-953.

27. Iversen, O.H., Urethane (ethyl carbamate) alone is carcinogenic for mouse skin, Carcinogenesis 5 (1984), 911-916.

28. Kendall, D.G., Birth and death processes and the theory of carcinogenesis, Biometrika **47** (1960), 316-330.

29. Klein, G., The approaching era of the tumor suppressor genes, Science **238** (1987), 1539-1545.

30. Klein, G. and Klein, E., Oncogene activation and tumor progression, Carcinogenesis 4 (1984), 429-436.

31. Klein, G. and Klein, E., Evolution of tumors and the impact of molecular oncology, Nature **315** (1985), 190-195.

32. Knudson, A.G., Mutation and cancer: Statistical study of retinoblastoma, Proc Natl. Acad. Sci. USA **68** (1971), 820-823.

33. Knudson, A.G., Hereditary cancer, oncogenes and antioncogenes, Cancer Res. **45** (1985), 1437-1443.

34. Koufos, A. et al., Loss of alleles at loci on human chromosome 11 during genesis of Wilm's tumor, Nature **309** (1984), 170-172.

35. Krailo, M., Thomas, D. and Pike, M., Fitting models of carcinogenesis to a case-control study of breast cancer, Symposium on "Time-Related Factors in Cancer Epidemiology", April 15-17, 1985. NCI/NIH, Bethesda, MD, Jour. Chronic Disease 40 Supplement 2 (1987).

36. Land, H., Parada, L.F., and Weinberg, R.A., Tumorigenic conversion of primary embryo fibroblasts requires at least two cooperating oncogenes, Nature **304** (1983), 596-601.

37. Lasnitski, I. and Bollag, W., Prevention and reversal by a retinoid of 3,4-benzpyrene and cigarette condensate induced hyperplasia and metaplasia of rodent respiratory epithelia in organ culture., Cancer Treatment Rep. **66** (1982), 1375-1380.

38. Mackillop, W.J., Ciampi, A. and Buck, R.N., A stem cell model of human tumor growth: Implications for tumor cell clonogenic assays, J. Nat. Cancer Inst. **70** (1983), 9-16.

39. MacMahon, B. et al., Age at first birth and breast cancer risk, Bull WHO **43** (1970), 209-221.

40. Marshall, C.J. and Ridby, P.W.J.C, Viral and cellular genes involved in oncogenesis, Cancer Survey **3** (1984), 183-214.

41. Matsumura, T., Hayashi, M. and Konishi, R., Immortalization in culture of rat cells: A genealogic study, J. Nat. Cancer Inst. **74** (1985), 1223-1232.

42. Marx, J.L., The Yin and Yang of cell growth control, Science **232** (1986), 1093- 1095.

43. McKnight, B. and Crowley, J., Tests for differences in tumor incidence based on animal carcinogenesis experiments, Jour. Amer. Statist. Assoc. **79** (1984), 639-648.

44. Moolgavkar, S.H., Model for human carcinogenesis: Action of environmental agents, Envir. Heath Persp. **50** (1983), 285- 291.

45. Moolgavkar, S.H., Carcinogenesis modeling: From molecular biology to epidemiology, Ann. Rev. Public Health **7** (1986), 151-169.

46. Moolgavkar, S.H., Hormones and multistage carcinogenesis, Cancer Survey **3** (1986), 183-214.

47. Moolgavkar, S.H. and Venzon, D.J., Two event model for carcinogenesis: Incidence curves for childhood and adult cancer, Math. Biosciences **47** (1979), 55-77.

48. Moolgavkar, S.H., Day, N.E. and Stevens, R.G., Two-stage model for carcinogenesis: Epidemiology of breast cancer in females, Jour. Nat. Cancer Inst. **65** (1980), 559-569.

49. Moolgavkar, S.H. and Knudson, A.G., Mutation and cancer: A model for human carcinogenesis, Jour. Nat. Cancer Inst. **66** (1981), 1037-1052.

50. Moolgavkar, S.H., Dewanji, A. and Venzon, D.J., A Stochastic two-stage model for cancer risk assessment I: The hazard function and the probability of tumor, Risk Analysis **3** (1988), 383-392.

51. Moolgavkar, S.H. and Luebeck, E.G., A biologically motivated parametric model for the analysis of animal tumorigenicity data, Fred

Hutchinson Cancer research Center. The University of Washington, Seattle, WA, 1989.

52. Neyman, J. and Scott, E., Statistical aspects of the problem of carcinogenesis. In: "Fifth Berkeley Symposium on Mathematical Statistics and Probability",, 745- 776, University of California Press, Berkeley, CA, 1967.

53. Oberley, L.W. and Oberley, T.D., The role of superoxide dismutase and gene amplification in carcinogenesis, J. Theor. Biology **106** (1984), 403-422.

54. Orkin, S.H. et al., Development of homozygosity for chromosome 11p markers in Wilm's tumors, Nature **309** (1984), 172-174.

55. Paterson, M.C. et al., Radiogenic neoplasia, cellular radiosensitivity and faulty DNA repair. In: "Radiation Carcinogenesis: Epidemiology and Biological Significance", 319-336, Raven Press, New York, 1984.

56. Portier, C., Estimating the tumor onset distribution in animal carcinogenesis experiments, Biometrika **73** (1986), 371-378.

57. Portier, C. and Dinse, G.E., Semiparametric analysis of tumor incidence rates in survival/sacrifice experiments, Biometrics **43** (1987), 107-114.

58. Portier, C.J. and Bailer, A.J., Two-stage models of tumor incidence for historical control animals in the national toxicology program's carcinogenicity experiments, NIEHS report, Research Triangle Park, NC 27709, 1989.

59. Portier, C.J., Hedges, J.C. and Hoel, D.G., Age-specific models of mortality and tumor onset for historical control animals in the national toxicology program's carcinogenicity experiments, Cancer Res. **46** (1986), 4372-4378.

60. Potter, V., Initiation and promotion in cancer formation: The importance of studies on intercellular communications, Yale J. Biol. Med. **53** (1980), 367-384.

61. Reitsma, P.H. et al., Regulation of myc gene expression in HL60 Leukemia cells by vitamin D metabolite, Nature **306** (1983), 492-494.

62. Scherer, E. et al., Initiation-promotion-initiation. Induction of neoplastic foci within islands of precancerous liver cells in the rat. In: "IARC Scientific Publication" No. **56** (1984), Lyon, France.

63. Shklar, G. et al., Regression by vitamin E of experimental oral cancer, Jour. Nat. Cancer Inst. **78** (1984), 987-992.

64. Sirover, M.A. et al., Cellular and Molecular regulation of DNA repair in normal human cells and in hypermutable cells from cancer prone

individual. In: "Proceeding of International Conference on Mechanisms of DNA Damage and Repair, June 2–7, 1985", edited by M.G. Simic, L. Grossman and A.C. Upton, Plenum Press, NY, 1986.

65. Solomon, E. et al., Chromosome 5 allele loss in human colorectal carcinomas, Nature **328** (1987), 616-619.

66. Strauss, B.S., Cellular aspects of DNA repair, Advances in Cancer Res. **45** (1985), 45-104.

67. Tan, W.Y., A stochastic Gompertz birth-death process, Statist and Prob. Lett. **4** (1986), 25-28.

68. Tan, W.Y. and Gastardo, M.T., On the assessment of effects of environmental agents on cancer tumor development by a two-stage model of carcinogenesis, Math Biosciences **74** (1985), 143-155.

69. Tan, W.Y. and Brown, C.C., A nonhomogeneous two-stage model of carcinogenesis, Math Modelling **9** (1987), 631-642.

70. Tan, W.Y. and Singh, K., On assessing the effects of metabolism of environmental agents on cancer tumor developments by a two-stage model of carcinogenesis, Env. Health Prespective **74** (1987), 203-210.

71. Tan, W.Y. and Piantadosi, S., On stochastic growth processes with application to stochastic logistic growth, To appear in Statistica Sinica, 1991.

72. Tan, W.Y. and Singh, K., A mixed model of carcinogenosis - with application to retinoblastoma, Math Bioscience **98** (1990), 201-211.

73. Tan, W.Y. and Starr, T., Estimating the times of initiation onset and tumor onset in animal carcinogenicty experiments with sacrifice by a two-stage model of carcinogenesis, Paper presented at the ASA-Biometric-IMS meeting, March 19-22, 1989.

74. Thorslund, T.W., Brown C.C. and Charnley, G., Biologically motivated cancer risk models, Risk Analysis **7** (1987), 109-119.

75. Topal, M.D., DNA repair, oncogenes and carcinogenesis, Carcinogenesis **9** (1988), 691-696.

76. Van Ryzin, J., Quantitative risk assessment, Jour. Occup. Medicine **22** (1980), 321-326.

77. Van Ryzin, J., Low dose assessment for regulating carcinogens, Center for the Social Sciences Newsletter, Columbia University, NY **4**, No. 1 (1984), 1-5.

78. Weinberg, R.A., Oncogenes, antioncogenes and the molecular bases of multistep carcinogenesis, Cancer Res. **49** (1989), 3713-3721.

79. Yuspa, S.H., Cellular and molecular changes during chemical carcinogenesis in mouse skin cells. In: "Carcinogenesis 10", 201-210, edited by E. Huberman and S.H. Barr, Raven Press, New York, 1985.

80. Zarbl, H., Sukumar, S. and Barbacid, M., Malignant activation of ras oncogenes in carcinogen-induced tumors, In: "Proceeding of International Conference on Mechanisms of DNA Damage and Repair", edited by M.G. Simic, L. Grossman and A.C. Upton, Plenum Press, NY, 1986.

4

Multiple Pathway Models of Carcinogenesis

As demonstrated in Chapter One, cancer tumors develop from a single normal stem cell by a multistep random process. In most of the literature, only a single pathway has been considered. In many practical situations, however, some cancers appear to have arisen from multiple pathways (2, 12, 16, 30). These observations have led Tan and Brown (50), Portier (39), and Tan and Chen (51) to develop multiple pathway models for carcinogenesis.

In Section (4.1) some biological evidence supporting multiple pathway models, especially multiple pathway models involving one-stage models and two-stage models, are presented. In Section (4.2) a multiple pathway model involving a one-stage model and two two-stage models is proposed. By using the results of Section (4.2), formula for age-specific incidence functions for the proposed multiple pathway model are derived in Section (4.3) while the expected number of tumors and the variances and the covariances of tumors are given in Section (4.4). The next three sections are devoted to some important special cases of multiple pathway models. Finally in Section (4.8), some possible applications of the multiple pathway models are indicated.

4.1 SOME BIOLOGICAL EVIDENCE

Foulds (12) appeared to be the first to have emphasized the importance of multiple pathway carcinogenesis processes. Originally, the necessity of considering multiple pathways for cancer development arose from the search for the explanation of inconsistent data from different research reports. For illustration, consider the melanoma development of skin cancer in human beings. Using the electron microscope, Mishima (31) found cellular differentiation between different types of melanoma cells, the HMF (Hutchinson's melanotic freckle) type, the SSM (superficial spreading melanoma) type and the UCM (unclassified type melanoma) type. Holman et al. (16) noted that these types of melanomas differed from each other in shapes of incidence curves, in frequency of occurrence, in different body sites and in reaction to ultraviolet radiation. These observations led Holman et al. (16) to propose a multiple pathway model involving two pathways for modular melanoma as shown in Figure 4.1 (taken from reference (16)). This two-pathway model provides a satisfactory explanation for many inconsistent reports on the relationships between skin cancer and sunlight.

Careful examination of the literature of cancer biology would reveal that multiple pathways for cancer might be quite common in real life (see references 2, 30). Given below are some specific biological evidence for multiple pathway processes of carcinogenesis.

Multiple Pathway Models of Carcinogenesis Involving One-Stage and Two-Stage Models

For some cancers, cancer tumors may develop either from an one-stage model or from a two-stage model; see Figure 4.2. Given below are some specific examples:

1. As demonstrated in Chapters One and Three, while in most of the cases carcinogenesis would require two stages, there are rare occasions in which one stage is enough to induce tumorigenic conversion of normal stem cells. For example, Spandidos and Wilkie (42) have shown that tumorigenic conversion of normal rodent cells can be achieved by a single activated human oncogene (ras) if it is linked to a strong promoter; also, as reported by Keath et al (19), if the normal neighboring cells are killed or removed by cytotoxic drugs, then activation of one oncogene in some cases also seems to be sufficient for inducing tumorigenic conversion of

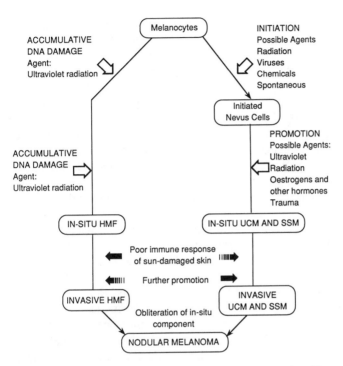

Figure 4.1 Proposed theory of the pathogenesis of human cutaneous malignant melanoma. (Reproduced with permission from reference (16).)

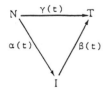

Figure 4.2 A multiple pathway model involving a one-stage model and a two-stage model of carcinogenesis. N = normal stem cell, I = intermediate cell, T = cancer tumor.

normal stem cells. Similar results have also been reported by Land et al (23), Detto, Weinberg and Ariza (10), and Bignami et al (4).

2. The existence of dual oncogenes implies that there are rare occassions in which one stage is sufficient to induce tumorigenic conversion of normal stem cells. For example, the SV40 large T antigen oncogene is known as a dual oncogene (6) having both immortalization and transformation functions; by the function of a single protein encoded by the SV40 large T antigen oncogene, the SV40 large T antigen oncogene alone is capable of inducing conversion to malignant tumor cells of normal stem cells (6, 24). Since for cancers related to oncogenes two stages are usually required for carcinogenesis in most of the cases (35, 47), the existence of dual oncogenes implies a multiple pathway model as shown in Figure 4.2.

3. Carcinogenesis involving one-stage models and two-stage models may also arise from situations in which the tissue in question at the birth time t_0 is a mixture of normal stem cells and initiated cells. As an illustration, consider cancers such as retinoblastoma or Wilm's tumor which are related to antioncogenes (20, 55). For these cancers, the Moolgavkar-Venzon-Knudson two-stage model provides a mathematical description of the biological mechanism (34, 35, 46, 47); further, the first event occurs either in germline cells or in somatic cells while the second event always occurs in somatic cells. If the first event occurs in somatic cells but before birth t_0, then at the time t_0 of birth the retina tissue consists of both initiated cells and normal stem cells. For these individuals, carcinogenesis develops either from initiated cells which involve a one-stage model, or from normal stem cells which involve two-stage models.

Multiple Pathway Models of Carcinogenesis Involving Two-Stage Models

In Chapters One and Three, it has been demonstrated that under usual conditions, tumorigenic conversion of normal stem cells can be achieved by cotransfection of one oncogene from the immortalization class with one oncogene from the transformation class (9, 14, 22, 28, 29, 54, 56, 57, 58). Conceivably, since oncogenes from the same class share common functions, one would expect that any oncogene from the immortalization class would collaborate with any oncogene from the transformation class to induce tumorigenic conversion of normal stem cells. Thus, the role of altered ras oncogene can be filled by the oncogene of the polyoma virus, and the role of the altered myc oncogene can be filled by the polyoma oncogene for large T antigen or by the ElA oncogene of adenovirus

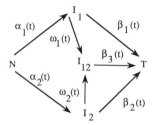

Figure 4.3 A multiple pathway model involving two-stage models of carcinogenesis. N = normal stem cell, I_j = type-j intermediate cell, $(I_{12} = I_3)$; T = cancer tumor cell.

5. In fact, it has been observed that the nuclear oncogenes myc and myb act synergistically with the cytoplasmic oncogenes src, erbB, fes/fps, yes, ros and mil/raf (1, 3); the viral oncogene E1A and polyoma and the cellular oncogenes myc, N-myc and P53 all collaborate with the cytoplasmic oncogene ras (22, 40, 58). These results suggest a multiple pathway model as shown in Figure 4.3.

Multiple pathway models as shown in Figure 4.3 have also been proposed by Medina (30) for cancers of mouse mammary gland. As reported by Medina (30), normal stem cells may develop neoplasms either through the pathway, Normal → HAN → Neoplasm, or through the pathway, Normal → DH → Neoplasm, where HAN=hyperplastic alveolar nodules and DH= ductal hyperplasias; further, neoplasms may also develop by the three-step pathways: Normal → HAN → HOG → Neoplasm and Normal → DH → HOG → Neoplasm, where HOG=preneoplastic hyperplastic outgrowth lines.

4.2 A MULTIPLE PATHWAY MODEL INVOLVING A ONE-STAGE MODEL AND TWO TWO-STAGE MODELS OF CARCINOGENESIS

From the previous biological evidence, I thus consider a general multiple pathway model as given in Figure 4.4 involving a one-stage model and two two-stage models.

In this model, a cancer tumor develops either by a one-stage pathway $(N \rightarrow T)$, or by any of the two two-step pathways $(N \rightarrow I_1 \rightarrow T)$ and

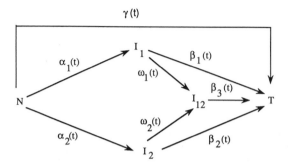

Figure 4.4 A multiple pathway model involving one-stage and two-stage models of carcinogenesis. N = normal stem cell, $I_1 = I_1$ cell, $I_2 = I_2$ cell, I_{12} (= I_3) = I_{12} cell, T = cancer tumor

$(N \rightarrow I_2 \rightarrow T)$, or by any of the two three-step pathways $(N \rightarrow I_1 \rightarrow I_{12} \rightarrow T)$ and $(N \rightarrow I_2 \rightarrow I_{12} \rightarrow T)$. Note that since the mutation rates are usually very small, in order that the two-step pathways and the three-step pathways are significant, the one-stage pathway should be a rare event which is in fact the case in many practical situations. Also, in order that the three-step pathways are important, the cell proliferation rate of I_{12} cells should be significantly greater than the cell proliferation rates of the I_i cells, i=1, 2. (For convenience of notation, in what follows we shall write I_3 for I_{12} unless otherwise stated.)

As illustrated in Chapter 6, in some cases carcinogenesis may involve three or more stages. However, since mutation rates are usually very small, carcinogenesis pathways involving three or more stages would usually be dominated by one-stage and two-stage pathways. As close approximations, in this chapter we thus only consider multiple pathway models of carcinogenesis involving one-stage models and two-stage models.

4.2.1 Some Basic Assumptions

Let $N(t), I_1(t), I_2(t), I_3(t)$ and $T(t)$ be the numbers of normal stem cells, I_1 cells, I_2 cells, $I_{12}(= I_3)$ cells and cancer tumors at time t, respectively. To derive mathematical theories for the model given in Figure 4.4, we make the following assumptions. Note that most of the assumptions are in fact the same assumptions for the two-stage models given in Chapter Three.

1. At the time t_0 of birth, $N(t_0) = N_0$ is very large ($N_0 \geq 10^6$). This assumption usually holds for most of the tissues; see references (35) and (47).

2. The I_j cells follow a nonhomogeneous Feller-Arley birth-death process for cell proliferation and cell differentiation with birth rate $b_j(t)$ and death rate $d_j(t)$. Note that if $b_j(t) - d_j(t) = \epsilon_j \exp(-\delta_j t)$ for some constants $\epsilon_j > 0$ and $\delta_j > 0$, the above birth-death process is then a stochastic Gompertz birth-death process (45).

3. Given that cancer tumors develop from normal stem cells by the one-stage pathway, we assume that the probability that a normal stem cell at time t yields one normal stem cell and one tumor cell at time $t + \Delta t$ is $\bar{\gamma}(t)\Delta t + o(\Delta t)$, where $\lim_{\Delta t \to 0} o(\Delta t)/\Delta t = 0$. Similarly, given the two-stage pathway $N \to I_i \to T$, we assume that the probability that a normal stem cell at time t yields one normal stem cell and one I_i cell at time $t + \Delta t$ is $\bar{\alpha}_i(t)\Delta t + o(\Delta t), i = 1, 2$. In what follows, we let $\gamma(t) = p_o(t)\bar{\gamma}(t)$ and $\alpha_i(t) = p_i(t)\bar{\alpha}_i(t), i = 1, 2$, where $p_o(t), p_1(t)$ and $p_2(t)$ are the probabilities for the pathways $N \to T, N \to I_1$ and $N \to I_2$ at time t, respectively. (Note $p_o(t) + p_1(t) + p_2(t) = 1$.)

4. The probability that an I_j cell at time t yields one I_j cell and one tumor cell at time $t + \Delta t$ is $\beta_j(t)\Delta t + o(\Delta t), j = 1, 2, 3$; the probability that an I_i cell at time t yields one I_i cell and one $I_{12}(= I_3)$ cell at time $t + \Delta t$ is $\omega_i(t)\Delta t + o(\Delta t), i = 1, 2$.

REMARK In the pathway $I_j \to T$, it may be either due to a one-stage pathway or due to the activation of an oncogene from the transformation class. Hence, $\beta_j(t)$ is in general a weighted sum of rates of the one-stage pathway and rates of activation of oncogenes from the transformation class.

5. As in Moolgavkar and Knudson (35) and Tan and Brown (47), we assume that the time required for the development of tumors from tumor cells is very short compared with the time required for the conversion of normal stem cells into cancer tumors. This implies that with probability one tumor cells grow into tumors and that one may ignore random variation for the time elapsed between the birth of tumor cells and the development of cancer tumors from tumor cells. Further, for human beings, the observed incidence rates are normally recorded for at least one

year; in fitting incidence curves for human cancers, one may thus ignore the time elapsed between the birth of tumor cells and the development of cancer tumors from tumor cells; see references (35) and (47).

6. The birth-death processes and the mutation processes are independent of one another and each cell goes through the above processes independently of other cells.

Given the above assumptions, the following results follow immediately:

1. The process is basically a Markov process. This follows from the fact that the future fate of any cell at time t depends only on the status of that cell at time t and is independent of past history of that cell.

2. Since the mutation rates are usually very small ($\approx 10^{-6}$–10^{-8}) and since in most of the cases the birth rates are greater than the death or differentration rates for normal stem cells (see 5, 25, 37), assumption (1) implies that $N(t)$ is very large for all $t \geq t_0$. In these cases, one may ignore random variations in $N(t)$ and assume deterministic growth for normal stem cells; see references (35) and (47). Further, since the mutation rates are very small so that $N(t)\gamma(t)$ and $N(t)\alpha_i(t), i = 1, 2$, are expected to be finite for all $t \geq t_0$. By the law of small numbers (limiting theory of binomial distribution), if $N(t)\gamma(t)$ and $N(t)\alpha_i(t), i = 1, 2$ are finite for all $t \geq t_0$, it is then appropriate to assume that during $[t, t + \Delta t]$, the mutation processes from normal stem cells to tumor cells, to I_1 cells and to I_2 cells follow independent Poisson processes with parameters $N(t)\gamma(t)\Delta t + o(\Delta t), N(t)\alpha_1(t)\Delta t + o(\Delta t)$ and $N(t)\alpha_2(t)\Delta t + o(\Delta t)$, respectively.

3. If the normal stem cells follow nonhomogeneous Feller-Arley birth-death processes for their proliferation and differentiation with birth rate $b_N(t)$ and death rate $d_N(t)$, then under the assumptions (2)–(6), $N(t), I_j(t), j = 1, 2, 3, T(t)$ form a continuous time multiple branching process. For these processes, the transition probabilities and hence the progeny distributions during $[t_j, t_j + \Delta t]$ are given in Table 4.1; see reference (47).

4.2.2 The Probabilities and the Probability Generating Functions (PGF)

Let $P_1(i, j_u, u = 1, 2, 3, k; t)$ be the conditional probability of $[N(t) = i, I_u(t) = j_u, u = 1, 2, 3, T(t) = k]$ given $N(t_0) = N_0, P_{2i}(j_i, j_3, k; t)$ the conditional probability of $[I_i(t) = j_i, I_3(t) = j_3, T(t) = k]$ given $I_i(t_0) = 1, i =$

Table 4.1 Transitions and Transition Probabilities of the Multiple Pathway Process Given in Figure 4.4.

Parent at t_j	Progenies at $t_j + \Delta t$	Probabilities
1 N Cell	2 N Cells	$b_N(t_j)\Delta t + o(\Delta t)$,
	0 Cells	$d_N(t_j)\Delta t + o(\Delta t)$,
	1 N Cell and 1 T Cell	$\gamma(t_j)\Delta t + o(\Delta t)$,
	1 N Cell and 1 I_1 Cell	$\alpha_1(t_j)\Delta t + o(\Delta t)$,
	1 N Cell and 1 I_2 Cell	$\alpha_2(t_j)\Delta t + o(\Delta t)$,
	1 N Cell	$1 - [b_N(t_j) + d_N(t_j) + \gamma(t_j) + \alpha_1(t_j) + \alpha_2(t_j)]\Delta t + o(\Delta t)$,
	Progenies other than above cases	$o(\Delta t)$
1 I_i Cell $i = 1, 2,$	2 I_i Cells	$b_i(t_j)\Delta t + o(\Delta t)$,
	0 Cells	$d_i(t_j)\Delta t + o(\Delta t)$,
	1 I_i Cell and 1 I_{12} Cell	$\omega_i(t_j)\Delta t + o(\Delta t)$
	1 I_i Cell and 1 T Cell	$\beta_i(t_j)\Delta t + o(\Delta t)$
	1 I_i cell	$1 - [b_i(t_j) + d_i(t_j) + \omega_i(t_j) + \beta_i(t_j)]\Delta t + o(\Delta t)$,
	Progenies other than above cases	$o(\Delta t)$
1 I_{12} ($= I_3$) Cell	2 I_{12} Cells	$b_3(t_j)\Delta t + o(\Delta t)$,
	0 Cells	$d_3(t_j)\Delta t + o(\Delta t)$,
	1 I_{12} Cell and 1 T Cell	$\beta_3(t_j)\Delta t + o(\Delta t)$,
	1 I_{12} Cell	$1 - [b_3(t_j) + d_3(t_j) + \beta_3(t_j)]\Delta t + o(\Delta t)$,
	Progenies other than above cases	$o(\Delta t)$
1 Tumor	1 Tumor	$1 + o(\Delta t)$
	Other cases	$o(\Delta t)$

$1, 2$ and $P_3(j,k;t)$ the conditional probability of $[I_3(t) = j, T(t) = k]$ given $I_3(t_0) = 1$.

Let $\eta(x,y_1,y_2,y_3,z;t_0,t) = \eta(t_0,t)$ be the PGF of $[N(t), I_1(t), I_2(t),$ $I_3(t), T(t)]$ given $N(t_0) = N_0; \eta(1,y_1,y_2,y_3,z;t_0,t) = \psi(y_1,y_2,y_3,z;t_0,t) = \psi(t_0,t)$ the PGF of $[I_1(t),I_2(t), I_3(t),T(t)]$ given $N(t_0) = N_0$ and $g(z;t_0,t) = \psi(1,1,1,z;t_0,t)$ the PGF of $T(t)$ given $N(t_0) = N_0$. Let $\zeta_i(y_i,y_3,z;t_0,t) = \zeta_i(t_0,t)$ be the PGF of $I_i(t),I_3(t)$ and $T(t)$ given one I_i cell at time $t_0, i = 1,2$; and $\phi(y_3,z;t_0,t) = \phi(t_0,t)$ the PGF of $I_3(t)$ and $T(t)$ given one $I_3(= I_{12})$ cell at time t_0.

Then,

$$\eta(t_0,t) = \sum_i \sum_{j_1} \sum_{j_2} \sum_{j_3} \sum_k x^i y_1^{j_1} y_2^{j_2} y_3^{j_3} z^k P_1(i,j_1,j_2,j_3,k;t),$$

$$\zeta_i(t_0,t) = \sum_{j_i} \sum_{j_3} \sum_k y_i^{j_i} y_3^{j_3} z^k P_{2i}(j_i,j_3,k;t), i = 1, 2,$$

and

$$\phi(t_0,t) = \sum_j \sum_k y^j z^k P_3(j,k;t).$$

Since the processes are Markov processes, one may then write down the Kolmogorov forward equations for the above conditional probabilities by using standard procedures. Using the transition probabilities given in Table (4.1), one has in fact the following Kolmogorov forward equations for $P_3(j,k;t)$ and $P_{2i}(j_i,j_3,k;t)$:

$$\frac{d}{dt}P_3(j,k;t) = (j - 1)b_3(t)P_3(j - 1,k;t)$$

$$+ (j + 1)d_3(t)P_3(j + 1,k;t) + j\beta_3(t)P_3(j,k - 1;t)$$

$$- j[b_3(t) + d_3(t) + \beta_3(t)]P_3(j,k;t), P_3(j,k;t_0) = \delta_{ij};$$

$$j,k = 0,1,2,\ldots; \tag{4.2.1}$$

$$\frac{d}{dt}P_{2i}(j_i,j_3,k;t) = (j_i - 1)b_i(t)P_{2i}(j_i - 1,j_3,k;t)$$

$$+ (j_i + 1)d_i(t)P_{2i}(j_i + 1,j_3,k;t)$$

$$+ j_i\,\omega_i(t)P_{2i}(j_i,j_3 - 1,k;t) + j_i\,\beta_i(t)P_{2i}(j_i,j_3,k - 1;t)$$

$$+ (j_3 - 1)b_3(t)P_{2i}(j_i,j_3 - 1,k;t) + (j_3 + 1)d_3(t)P_{2i}(j_i,j_3 + 1,k;t)$$

$$+ j_3\,\beta_3(t)P_{2i}(j_i,j_3,k - 1;t) - \{j_i\,[b_i(t) + d_i(t) + \omega_i(t)] + \beta_i(t)$$

$$+ j_3\,[b_3(t) + d_3(t) + \beta_3(t)]\}P_{2i}(j_i,j_3,k;t), P_{2i}(j_i,j_3,k;t_0)$$

$$= \delta_{1j_i},j_i,j_3,k = 0,1,2,\ldots; \tag{4.2.2}$$

if the normal stem cells follow a non-homogeneous Feller-Arley birth-death process for cell proliferation and cell differentiation with birth rate $b_N(t)$ and death rate $d_N(t)$, then $P_1(i, j_u, u = 1, 2, 3, k; t)$ satisfies the following Kolmogorov forward equation:

$$\frac{d}{dt} P_1(i, j_1, j_2, j_3, k; t)$$
$$= (i - 1)b_N(t)P_1(i - 1, j_1, j_2, j_3, k; t) + (i + 1)d_N(t)$$
$$\times P_1(i + 1, j_1, j_2, j_3, k; t)$$
$$+ \sum_{u=1}^{3} \{(j_u - 1)b_u(t)P_1(i, j_u - 1, j_v, v \neq u, v = 1, 2, 3, k; t)$$
$$+ (j_u + 1)d_u(t)P_1(i, j_u + 1, j_v, v \neq u, v = 1, 2, 3, k; t)\}$$
$$+ \sum_{u=1}^{2} \{i\, \alpha_u(t)P_1(i, j_u - 1, j_v, v \neq u, v = 1, 2, j_3, k; t)$$
$$+ j_u\, \omega_u(t)P_1(i, j_1, j_2, j_3 - 1, k; t)\}$$
$$+ [i\, \gamma(t) + \sum_{u=1}^{3} j_u\, \beta_u(t)]P_1(i, j_1, j_2, j_3, k - 1; t)$$
$$- \{i\, [b_N(t) + d_N(t) + \gamma(t) + \alpha_1(t)$$
$$+ \alpha_2(t)] + \sum_{u=1}^{2} j_u\, \omega_u(t) + \sum_{u=1}^{3} j_u[b_u(t) + d_u(t) + \beta_u(t)]\} \qquad (4.2.3)$$
$$P_1(i, j_1, j_2, j_3, k; t), P_1(i, j_1, j_2, j_3, k; t_0) = \delta_{iN_0}, i, j_1, j_2, j_3, k = 0, 1, 2, \dots,$$

Using the above Kolmogorov forward equations, we have the following theorem for the above PGF's.

THEOREM (4.2.1) Under the assumptions (2)–(6), $\zeta_i(t_0, t), i = 1, 2,$ and $\phi(t_0, t)$ satisfy, respectively, the following first order partial differential equations:

$$\frac{\partial}{\partial t}\, \zeta_i(t_0, t) = \{(y_i - 1)[y_i b_i(t) - d_i(t)] + y_i(y_3 - 1)\omega_i(t)$$
$$+ y_i(z - 1)\beta_i(t)\} \frac{\partial}{\partial y_i}\, \zeta_i(t_0, t) + \{(y_3 - 1)[y_3 b_3(t) - d_3(t)]$$
$$+ y_3(z - 1)\beta_3(t)\} \frac{\partial}{\partial y_3}\, \zeta_i(t_0, t), \zeta_i(t_0, t_0) = y_i, i = 1, 2;$$

$$(4.2.4)$$

$$\frac{\partial}{\partial t}\phi(t_0,t) = \{(y_3 - 1)[y_3 b_3(t) - d_3(t)] + y_3(z - 1)\beta_3(t)\}$$

$$\times \frac{\partial}{\partial y_3}\phi(t_0,t), \phi(t_0,t_0) = y_3. \qquad (4.2.5)$$

If the normal stem cells follow a nonhomogeneous Feller-Arley birth-death process with birth rate $b_N(t)$ and death rate $d_N(t)$ for cell proliferation and differentiation, then $\eta(t_0,t)$ also satisfies the following first order partial differential equation:

$$\frac{\partial}{\partial t}\eta(t_0,t) = \{(x - 1)[xb_N(t) - d_N(t)] + x(z - 1)\gamma(t)$$

$$+ \sum_{i=1}^{2} x(y_i - 1)\alpha_i(t)\} \frac{\partial}{\partial x}\eta(t_0,t) + \sum_{i=1}^{2}\{(y_i - 1)[y_i b_i(t) - d_i(t)]$$

$$+ y_i(z - 1)\beta_i(t) + y_i(y_3 - 1)\omega_i(t)\} \frac{\partial}{\partial y_i}\eta(t_0,t)$$

$$+ \{(y_3 - 1)[y_3 b_3(t) - d_3(t)] + y_3(z - 1)\beta_3(t)\} \frac{\partial}{\partial y_3}\eta(t_0,t), \eta(t_0,t_0) = x^{N_0};$$

$$(4).2.6$$

Proof. By using the Kolmogorov forward equations (4.2.1)–(4.2.3), the proof is almost exactly the same as proving (3.2.3)–(3.2.5) of Chapter Three for two-stage models and is very Straightforward.

As we have demonstrated in Chapter Three, solution of equation (4.2.5) is available for most of the important special cases. However, solutions of equations (4.2.4) and (4.2.6) are extremely difficult even under homogeneous cases. Nevertheless, one may use equations (4.2.4) and (4.2.6) to obtain moments of the number of tumors, in particular the mean and the variance of the number of tumors. Furthermore, if $N(t)$ is very large but $\gamma(t)$ and $\alpha_i(t)$ are very small so that $N(t)\gamma(t)$ and $N(t)\alpha_i(t)$ are finite for all $t \geq t_0, i = 1, 2$, by the law of small numbers, one may then assume Poisson distributions for the number of mutations from normal stem cells to tumor cells and to I_i cells, $i = 1, 2$. This leads to the following theorem for $\psi(t_0,t)$.

THEOREM (4.2.2)   Assume that $N(t)\gamma(t)$ and $N(t)\alpha_i(t)$ are finite for all $t \geq t_0, i = 1, 2$, and that during $[t, t + \Delta t]$, the numbers of mutations from normal stem cells to I_i cells by the two-stage pathways, $i = 1, 2$

and from normal stem cells to T cells by the one-stage pathway follow independent Poisson distributions with parameters $N(t)\alpha_1(t)\Delta t + o(\Delta t)$, $N(t)\alpha_2(t)\Delta t + o(\Delta t)$ and $N(t)\gamma(t)\Delta t + o(\Delta t)$, respectively.

Then, under assumptions (1)–(6), $\psi(t_0,t)$ is given by:

$$\psi(t_0,t) = \exp\left\{(z-1)\int_{t_0}^{t} N(u)\gamma(u)du\right.$$

$$\left. + \sum_{j=1}^{2}\int_{t_0}^{t} N(u)\alpha_j(u)[\zeta_j(u,t) - 1]du\right\}, \tag{4.2.7}$$

where $N(u)$ is the expected number of normal stem cells at time $t = u$ given $N(t_0) = N_0$.

Proof. Partition the time interval $[t_0,t]$ by $L_j = [t_{j-1},t_j), j = 1,2,\ldots,n-1$, and $L_n = [t_{n-1},t_n]$, where $t_j = t_0 + j\Delta t$ and $n\Delta t = t_n - t_0$ with $t_n = t$. Let M_{0j}, M_{1j} and M_{2j} be the number of mutations from N to T cells, from N to I_1 cells and from N to I_2 cells, respectively, in $L_j, j = 1,\ldots,n$. Then the conditional PGF of $I_u(t), u = 1,2,3$ and $T(t)$ given the partition and given $(M_{0j},M_{1j},M_{2j}, j = 1,\ldots,n)$ is

$$\xi(t;\Delta t) = \prod_{j=1}^{n}\{z^{M_{0j}}[\zeta_1(t_{j-1},t)]^{M_{1j}}[\zeta_2(t_{j-1},t)]^{M_{2j}}\}.$$

Since M_{0j}, M_{1j} and M_{2j} follow Poisson distributions independently with parameters $N(t_{j-1})\gamma(t_{j-1})\Delta t + o(\Delta t), N(t_{j-1})\alpha_1(t_{j-1})\Delta t + o(\Delta t)$ and $N(t_{j-1})\alpha_2(t_{j-1})\Delta t + o(\Delta t)$, respectively, by taking expectation over $[M_{0j},M_{1j},M_{2j}, j = 1,\ldots,n]$, one obtains the conditional PGF $Qp(t_0,t;\Delta t)$ of $I_u(t), u = 1,2,3$ and $T(t)$ given $N(t_0) = N_0$ and given the partition $[L_j, j = 1,\ldots,n]$ of $[t_0,t]$ as

$$Qp(t_0,t;\Delta t) = \exp\left\{(z-1)\sum_{j=1}^{n}N(t_{j-1})\gamma(t_{j-1})\Delta t\right.$$

$$\left. + \sum_{i=1}^{2}\sum_{j=1}^{n}N(t_{j-1})\alpha_i(t_{j-1})[\zeta_i(t_{j-1},t) - 1]\Delta t + o(\Delta t)\right\}.$$

It follows that

$$\psi(t_0,t) = \lim_{\Delta t \to 0} Q_p(t_0,t;\Delta t) = \exp\left\{(z-1)\int_{t_0}^{t} N(x)\gamma(x)dx\right.$$

$$+ \sum_{i=1}^{2} \int_{t_0}^{t} N(x)\alpha_i(x)[\zeta_i(x,t) - 1]dx\}.$$

If the normal stem cells follow a nonhomogeneous Feller-Arley birth-death process with birth rate $b_N(t)$ and death rate $d_N(t)$, then one may alternatively derive formula (4.2.7) by using a continuous time multiple branching process. This result we give in the following corollary of Theorem (4.2.2).

COROLLARY   Assume that the normal stem cells follow a nonhomogeneous Feller-Arley birth-death process with birth rate $b_N(t)$ and death $d_N(t)$. If N_0 is very large but $\gamma(t)$ and $\alpha_i(t)$ very small for all $t \geq t_0$ so that $N_0\gamma(t)$ and $N_0\alpha_i(t)$ are finite for all $t \geq t_0$, then, under conditions (1)–(6), to order $O(N_0^{-1})$, $\psi(t_0,t)$ is given by (4.2.7).

Using the multiple branching process method, the proof of the above Corollary is almost exactly the same as the proof of Theorem (2) of Tan and Brown (47). The proof is quite straightforward although the notations are long and inconvenient; see Tan and Brown (47).

In (4.2.7), note that for the one-stage pathway, the PGF is that of a nonhomogeneous Poisson process while for the two-stage pathways, the PGF is that of a bivariate nonhomogeneous filtered Poisson process as defined in Tan (44).

To use (4.2.7) to obtain the incidence function of tumors and to obtain moments of the number of tumors, one would need the expected number $N(t)$ of normal stem cells and $\zeta_j(s,t)$. In Sections (4.5)–(4.7), we shall illustrate how to approximate $\zeta_j(s,t)$ under important special cases.

As we have demonstrated in Chapter Three, the functional form of $N(t)$ depends on different tissues.

1. For breast cancer of women, Moolgavkar and Knudson (35), and Moolgavkar (32, 33) showed that the growth of breast tissue cells of women is best described by logistic growth. Hence, given $N(t_0) = N_0$,

$$N(t) = N_0 \exp\{ \int_{t_0}^{t} \epsilon_N(x)dx \}$$

$$\times \{1 - (N_0/M) + (N_0/M) \exp [\int_{t_0}^{t} \epsilon_N(x)dx]\}^{-1},$$

    where $\epsilon_N(t) = b_N(t) - d_N(t)$.

2. For the lung and the colon, the growth of stem cells is best described by Gompertz growth (see 32). Hence,

$$N(t) = N_0 \exp\{(\epsilon_N/\delta)[\exp(-\delta t_0) - \exp(-\delta t)]\},$$

where $\epsilon_N > 0$ and $\delta > 0$.

3. For retinoblastoma, Wilm's tumor and Hodgkin's disease, the growth of stem cells is best fitted by Gamma curves or exponential curves (35, 48). Hence

$$N(t) = aN_0(t - t_0)^{c-1} \exp[-\epsilon_N(t - t_0)]$$

where $c \geq 1$, $\epsilon_N \geq 0$ and $a > 0$.

4.3 THE AGE-SPECIFIC INCIDENCE FUNCTIONS OF TUMORS FOR THE MODEL IN FIGURE 4.4

Let $\lambda(t)$ be the incidence rate of tumors at time t. Then, by Theorem (2.1.1) of Chapter Two,

$$\lambda(t) = -\psi'(1, 1, 1, 0; t_0, t)/\psi(1, 1, 1, 0; t_0, t),$$

where $\quad \psi'(1, 1, 1, 0; t_0, t) = d\psi(1, 1, 1, 0; t_0, t)/dt.$

Let $\psi(1, 1, 1, 0; t_0, t) = \psi_{10}(t_0, t)$. Since $\psi(y_1, y_2, y_3, z; t_0, t) = \psi(t_0, t)$, by Theorem (4.2.2), one has to the order $O(N_0^{-1})$:

$$\psi'(1, 1, 1, 0; t_0, t) = \psi(1, 1, 1, 0; t_0, t)\{-N(t)\gamma(t)$$

$$+ \sum_{i=1}^{2} N(t)\alpha_i(t)[\zeta_i(1, 1, 0; t, t) - 1]$$

$$+ \sum_{i=1}^{2} \int_{t_0}^{t} N(u)\alpha_i(u)\nu_i(u, t)du\},$$

where

$$\nu_i(u, t) = \partial\zeta_i(1, 1, 0; u, t)/\partial t.$$

Since $\zeta_i(y_i, y_3, z; t_0, t_0) = \zeta_i(t_0, t_0) = y_i$, so

$$\lambda(t) = -\psi'(1, 1, 1, 0; t_0, t)/\psi(1, 1, 1, 0; t_0, t)$$

$$= N(t)\gamma(t) + \sum_{i=1}^{2} \int_{t_0}^{t} N(u)\alpha_i(u)[-\nu_i(u, t)]du, \qquad (4.3.1)$$

Now, from equation (4.2.4),

$$-\nu_i(u,t) = \beta_i(t)[\partial\zeta_i(u,t)/\partial y_i]_{(y_i=y_3=1,z=0)}$$
$$+ \beta_3(t)[\partial\zeta_i(u,t)/\partial y_3]_{(y_i=y_3=1,z=0)}$$
$$= \beta_i(t)\zeta_{i0}(u,t)E[I_i(t)|T(t)=0,I_i(u)=1]$$
$$+ \beta_3(t)\zeta_{i0}(u,t)E[I_3(t)|T(t)=0,I_i(u)=1],$$

where $\zeta_{i0}(u,t) = \zeta_i(1,1,0;u,t)$ and $E[I_j(t)|T(t)=0,I_i(u)=1]$ is the conditional expected number of I_j cells at time t given $[T(t)=0,I_i(u)=1]$ for $t \geq u$.

It follows that

$$\lambda(t) = N(t)\gamma(t) + \sum_{i=1}^{2}\int_{t_0}^{t} N(u)\alpha_i(u)\zeta_{i0}(u,t)$$
$$\times [\beta_i(t)\mu_{i,i}(u,t) + \beta_3(t)\mu_{i,3}(u,t)]du, \qquad (4.3.2)$$

where

$$\mu_{i,j}(u,t) = E[I_j(t)|T(t)=0,I_i(u)=1].$$

4.3.1 An Approximation

To obtain $\lambda(t)$ by using (4.3.2), one would need the solution $\zeta_i(u,t)$ of equation (4.2.4). Since solution of (4.2.4) is extremely difficult, one needs to seek an approximation to $\lambda(t)$. One such approximation can be obtained by noting that the mutation rates $\beta_j(t)$ are very small ($\beta_i(t) \approx 10^{-6}$–$10^{-8}$) for all $t \geq t_0$. Now, if $\beta_j(t) = 0, j = 1,2,3$, for all $t \geq t_0$, then $\zeta_{10}(u,t) = 1$. Hence, if $\beta_j(t), j = 1,2,3$ is very small ($\beta_j(t) \approx 10^{-6}$–$10^{-8}$) for $t_1 \geq t \geq u$ and if $\epsilon_j(t), j = 1,2,3$ is not big ($\epsilon_j(t) \leq 0.1$) for $t_1 \geq t \geq u$, then for $u \leq t \leq t_1, \mu_{ij}(u,t)$ is closely approximated by:

$$\mu_{ij}(u,t) \cong m_{ij}(u,t) = E[I_j(t)\,|\,I_u(t)].$$

Taking the derivative of $\zeta_i(u,t)$ with respect to $y_j, j = i,3$ and putting $y_i = y_3 = z = 1$, one obtains from equation (4.2.4):

$$\frac{d}{dt}m_{i,i}(u,t) = \epsilon_i(t)m_{i,i}(u,t), \qquad (4.3.3)$$

$$\frac{d}{dt}m_{i,3}(u,t) = \omega_i(t)m_{ii}(u,t) + \epsilon_3(t)m_{i,3}(u,t), \qquad (4.3.4)$$

$$m_{i,i}(u,u) = 1, m_{i,3}(u,u) = 0, i = 1,2, \quad \text{where} \quad \epsilon_j(t) = b_j(t) - d_j(t).$$

From (4.3.3), one obtains, for $i = 1, 2$:

$$m_{i,i}(u,t) = \exp\{\int_u^t \epsilon_i(x)dx\},\tag{4.3.5}$$

From (4.3.4), one obtains, for $i = 1, 2$:

$$m_{i,3}(u,t) = \int_u^t \omega_i(y)m_{i,i}(u,y) \exp[\int_y^t \epsilon_3(x)dx]dy$$

$$= \int_u^t \omega_i(y) \exp\{\int_u^y \epsilon_i(x)dx + \int_y^t \epsilon_3(x)dx\}dy.\tag{4.3.6}$$

On substituting (4.3.5) and (4.3.6) into (4.3.2), if $\beta_j(t)$ is very small and if $\epsilon_j(t)$ is not big for all $t \geq t_0$ and for $j = 1, 2, 3$, an approximation to $\lambda(t)$ is given by:

$$\lambda(t) \cong N(t)\gamma(t) + \sum_{i=1}^2 \int_{t_0}^t N(u)\alpha_i(u) [\beta_i(t)m_{i,i}(u,t) + \beta_3(t)m_{i,3}(u,t)]du,$$

$$\tag{4.3.7}$$

where $m_{i,i}(u,t)$ and $m_{i,3}(u,t)$ are given respectively in (4.3.5) and (4.3.6).

4.4 THE EXPECTED VALUE AND THE VARIANCE OF THE NUMBER OF TUMORS FOR THE MODEL IN FIGURE 4.4

Given the PGF $\psi(t_0, t)$ in (4.2.7), theoretically one may derive the probabilities of the number of tumors. However, since the solution $\zeta_i(u,t)$ of equation (4.2.4) is very difficult to obtain in general cases, it is not an easy job to compute the probabilities of the number of tumors. In any case, one may always use (4.2.4) and (4.2.7) to obtain moments of the number of tumors. In particular, one may use (4.2.4) and (4.2.7) to obtain the expected value $u(t)$ and the variance $V(t)$ of the number of tumors given $N(t_0) = N_0$.

To obtain $u(t)$ and $V(t)$, we put:

$$m_{i,j}(s,t) = [\partial\zeta_i(s,t)/\partial y_j]_{(y_v=z=1, v=1,2,3)},$$

$$\mu_i(s,t) = [\partial\zeta_i(s,t)/\partial z]_{(y_v=z=1, v=1,2,3)};$$

$$V_i(s,t) = \text{Variance of } T(t) \text{ given } I_i(s) = 1,$$

$$C_{u,v}^{(i)}(s,t) = \text{Covariance of } I_u(t) \text{ and } I_v(t) \text{ given } I_i(s) = 1,$$

$$i = 1, 2, u, v = 1, 2, 3,$$

and

$$C_{u,T}^{(i)}(s,t) = \text{Covariance of } I_u(t) \text{ and } T(t) \text{ given } I_i(s) = 1,$$
$$i = 1, 2, u = 1, 2, 3.$$

Then, taking derivatives of (4.2.7) and setting $y_v = z = 1, v = 1, 2, 3$, one has:

$$u(t) = \int_{t_0}^{t} N(x)\gamma(x)dx + \sum_{i=1}^{2} \int_{t_0}^{t} N(x)\alpha_i(x)\mu_i(x,t)dx, \qquad (4.4.1)$$

and

$$V(t) = \int_{t_0}^{t} N(x)\gamma(x)dx + \sum_{i=1}^{2} \int_{t_0}^{t} N(x)\alpha_i(x)[V_i(x,t) + \mu_i^2(x,t)]dx, \quad (4.4.2)$$

4.4.1   Computing $\mu_i(x,t)$ and $V_i(x,t)$

To compute $u(t)$ by (4.4.1) and $V(t)$ by (4.4.2), one would need $\mu_i(x,t)$ and $V_i(x,t)$.

Taking derivatives of (4.2.4) and setting $y_v = z = 1, v = 1, 2, 3$, we obtain:

$$\frac{d}{dt}\mu_i(t_0,t) = \beta_i(t)m_{i,i}(t_0,t) + \beta_3(t)m_{i,3}(t_0,t), \quad \mu_i(t_0,t_0) = 0 \qquad (4.4.3)$$

where $m_{i,i}(t_0,t)$ and $m_{i,3}(t_0,t)$ are given in (4.3.5) and (4.3.6) respectively.
From (4.4.3)

$$\mu_i(t_0,t) = \int_{t_0}^{t} [\beta_i(x)m_{i,i}(t_0,x) + \beta_3(x)m_{i,3}(t_0,x)]dx, \qquad (4.4.4)$$

To evaluate $V_i(s,t)$, observe that

$$\{\partial^2 \zeta_i(s,t)/\partial y_u \partial y_v\}_{(y_j=z=1,j=1,2,3)}$$
$$= C_{u,v}^{(i)}(s,t) - \delta_{uv}m_{i,u}(s,t) + m_{i,u}(s,t)m_{i,v}(s,t).$$

Taking derivatives on both sides of equation (4.2.4) and setting $y_v = z = 1, v = 1, 2, 3$, we obtain the following system of linear differential equations for $V_i(s,t), C_{i,T}^{(i)}(s,t), C_{3,T}^{(i)}(s,t)$ and $C_{u,v}^{(i)}(s,t), i = 1, 2, u, v = 1, 2, 3$.

$$\frac{d}{dt}V_i(s,t) = \beta_i(t)[m_{i,i}(s,t) + 2C_{i,T}^{(i)}(s,t)] + \beta_3(t)[m_{i,3}(s,t)$$

$$+ 2C_{3,T}^{(i)}(s,t)],$$

$$V_i(s,s) = 0, \; i = 1,2; \tag{4.4.5}$$

$$\frac{d}{dt} C_{i,T}^{(i)}(s,t) = \beta_i(t)C_{i,i}^{(i)}(s,t) + \beta_3(t)C_{i,3}^{(i)}(s,t) + \epsilon_i(t)C_{i,T}^{(i)}(s,t),$$

$$C_{i,T}^{(i)}(s,s) = 0, i = 1,2, \tag{4.4.6}$$

$$\frac{d}{dt} C_{3,T}^{(i)}(s,t) = \beta_i(t)C_{i,3}^{(i)}(s,t) + \beta_3(t)C_{3,3}^{(i)}(s,t) + \omega_i(t)C_{i,T}^{(i)}(s,t)$$

$$+ \epsilon_3(t)C_{3,T}^{(i)}(s,t),$$

$$C_{3,T}^{(i)}(s,s) = 0, i = 1,2; \tag{4.4.7}$$

$$\frac{d}{dt} C_{i,i}^{(i)}(s,t) = [b_i(t) + d_i(t)]m_{i,i}(s,t) + 2\epsilon_i(t)C_{i,i}^{(i)}(s,t),$$

$$C_{i,i}^{(i)}(s,s) = 0, \; i = 1,2; \tag{4.4.8}$$

$$\frac{d}{dt} C_{i,3}^{(i)}(s,t) = [\epsilon_i(t) + \epsilon_3(t)]C_{i,3}^{(i)}(s,t) + \omega_i(t)C_{i,i}^{(i)}(s,t),$$

$$C_{i,3}^{(i)}(s,s) = 0, \; i = 1,2; \tag{4.4.9}$$

and

$$\frac{d}{dt} C_{3,3}^{(i)}(s,t) = [b_3(t) + d_3(t)]m_{i,3}(s,t) + \omega_i(t)m_{i,i}(s,t)$$

$$+ 2\omega_i(t)C_{i,3}^{(i)}(s,t)$$

$$+ 2\epsilon_3(t)C_{3,3}^{(i)}(s,t), C_{3,3}^{(i)}(s,s) = 0, i = 1,2. \tag{4.4.10}$$

The above system of differential equations can easily be solved iteratively as follows:

From (4.4.8), we obtain:

$$C_{i,i}^{(i)}(s,t) = \int_s^t [b_i(x) + d_i(x)]m_{i,i}(s,x) \exp [2 \int_x^t \epsilon_i(z)dz]dx, \tag{4.4.11}$$

From (4.4.9), we obtain:

$$C_{i,3}^{(i)}(s,t) = \int_s^t \omega_i(x)C_{i,i}^{(i)}(s,x) \exp \{ \int_x^t [\epsilon_i(u) + \epsilon_3(u)]du\}dx, \tag{4.4.12}$$

From (4.4.10), we obtain:

$$C_{3,3}^{(i)}(s,t) = \int_s^t \{[b_3(x) + d_3(x)]m_{i,3}(s,x) + \omega_i(x)m_{i,i}(s,x)$$

$$+2\omega_i(x)C_{i,3}^{(i)}(s,x)\}\exp\{2\int_x^t \epsilon_3(u)du\}dx,\tag{4.4.13}$$

From (4.4.6), we obtain:

$$C_{i,T}^{(i)}(s,t) = \int_s^t [\beta_i(x)C_{i,i}^{(i)}(s,x) + \beta_3(x)C_{i,3}^{(i)}(s,x)]\exp[\int_x^t \epsilon_i(u)du]dx,\tag{4.4.14}$$

From (4.4.7), we obtain:

$$C_{3,T}^{(i)}(s,t) = \int_s^t [\beta_i(x)C_{i,3}^{(i)}(s,x) + \beta_3(x)C_{3,3}^{(i)}(s,x)$$

$$+\omega_i(x)C_{i,T}^{(i)}(s,x)]\exp[\int_x^t \epsilon_3(u)du]dx,\tag{4.4.15}$$

Finally, from (4.4.5), we obtain:

$$V_i(s,t) = \int_s^t \{\beta_i(x)[m_{i,i}(s,x) + 2C_{i,T}^{(i)}(s,x)]$$

$$+\beta_3(x)[m_{i,3}(s,x) + 2C_{3,T}^{(i)}(s,x)]\}dx.\tag{4.4.16}$$

4.5 THE SEMI-HOMOGENEOUS MODEL

In the model given by Figure (4.4), if $\omega_i(t) = \omega_i, i = 1,2, \beta_j(t) = \beta_j, b_j(t) = b_j$ and $d_j(t) = d_j, j = 1,2,3$, are independent of t, then $\zeta_i(s,t) = \zeta_i(t-s), i = 1,2$ and $\phi(s,t) = \phi(t-s)$. In this case, (4.2.7) becomes

$$\psi(t_0,t) = \exp\{(z-1)\int_{t_0}^t N(u)\gamma(u)du + \sum_{i=1}^2 \int_{t_0}^t N(u)\alpha_i(u)$$

$$\times[\zeta_i(t-u)-1]du\}.\tag{4.5.1}$$

4.5.1 The Probability Generating Functions (PGF) of the Number of Cancer Tumors

Using exactly the same procedure as in proving Theorem (3.3.1) of Chapter Three, it is easily shown that $\zeta_i(t), i = 1,2$ and $\phi(t)$ satisfying, respectively, the following Ricatti equations:

$$\frac{d}{dt}\zeta_i(t) = b_i\zeta_i^2(t) + [\omega_i\phi(t) + \beta_i z - (b_i + d_i + \omega_i + \beta_i)]\zeta_i(t) + d_i,$$

$$\zeta_i(0) = y_i, i = 1,2;\tag{4.5.2}$$

$$\frac{d}{dt}\phi(t) = b_3\phi^2(t) + [\beta_3 z - (b_3 + d_3 + \beta_3)]\phi(t) + d_3, \phi(0) = y_3. \quad (4.5.3)$$

These results imply that if $\omega_i(t) = \omega_i, i = 1, 2, \beta_j(t) = \beta_j, b_j(t) = b_j$ and $d_j(t) = d_j, j = 1, 2, 3$, then $\phi(t)$ satisfies both equations (4.2.5) and (4.5.3) while $\zeta_i(t)$ satisfies both equations (4.2.4) and (4.5.2).

Solutions of (4.5.2) and (4.5.3)

Comparing equations (4.5.2) and (4.5.3) with equations (3.3.2) and (3.3.1) of Chapter Three, it is immediately seen that equations (4.5.2) and (4.5.3) are exactly the same equations (3.3.2) and (3.3.1) of Chapter Three, respectively. Hence, the solution of equation (4.5.3) is available from (3.3.3) and is given by:

$$\phi(t) = \phi(y_3, z; t)$$
$$= \{a_2(y_3 - a_1) + a_1(a_2 - y_3) \exp [(a_2 - a_1)b_3 t]\}$$
$$\times \{(y_3 - a_1) + (a_2 - y_3) \exp [(a_2 - a_1)b_3 t]\}^{-1}, \quad (4.5.4)$$

where $a_2 > a_1$ are defined by

$$2b_3 a_2 = (b_3 + d_3 + \beta_3 - z\beta_3) + h_3(z)$$

and

$$2b_3 a_1 = (b_3 + d_3 + \beta_3 - z\beta_3) - h_3(z)$$

with

$$h_3(z) = [(b_3 + d_3 + \beta_3 - z\beta_3)^2 - 4b_3 d_3]^{1/2}.$$

As we have shown in Chapter Three, the solution of equation (4.5.2) is in general quite difficult to obtain although a series solution may be derived for (4.5.2) by using a result given in Moolgavkar and Venzon (34). If $d_i = 0, i = 1, 2$, equation (4.5.2) then becomes

$$\frac{d}{dt}\zeta_i(t) = b_i\zeta_i^2(t) + [\omega_i\phi(t) + \beta_i z - (b_i + \omega_i + \beta_i)]\zeta_i(t), \zeta_i(0) = y_i. \quad (4.5.5)$$

Letting $g_i(t) = \zeta_i^{-1}(t)$ and dividing both sides of (4.5.5) by $\zeta_i^2(t)$, one obtains from (4.5.5):

$$-\frac{d}{dt}g_i(t) = b_i + g_i(t)[\omega_i\phi(t) + \beta_i z - (b_i + \omega_i + \beta_i)], g_i(0) = 1/y_i.$$

It follows that the solution of (4.5.5) is given by

$$\zeta_i(t) = y_i\theta_i(t)\{1 - y_i b_i U_i(t)\}^{-1}, \quad (4.5.6)$$

where

$$\theta_i(t) = \exp\{t(z-1)\beta_i - tb_i + \omega_i \int_0^t [\phi(x) - 1]dx\}$$

and

$$U_i(t) = \int_0^t \theta_i(y)dy.$$

Some Approximations of $\zeta_i(t)$

Since the mutation rates $\omega_i, i = 1, 2$, and $\beta_j, j = 1, 2, 3$ are usually very small ($\omega_i \approx 10^{-6}$–10^{-8}, $\beta_j \approx 10^{-6}$–10^{-8}) while the proliferation rates $\epsilon_j = b_j - d_j, j = 1, 2, 3$ are usually not big ($\epsilon_j \leq 0.1$), some close approximations to $\zeta_i(t)$ can readily be derived.

To obtain these approximations, observe from formula (4.5.4) that if $\beta_3 = 0$, then $h_3(z) = b_3 - d_3 = \epsilon_3$ so that $a_2 = 1$ and $a_1 = d_3/b_3$ ($b_3 \geq d_3$ when $\epsilon_3 \geq 0$). If follows that if $\beta_3 = 0$, then $\phi(t)$ is independent of z and is given by:

$$\phi(t) = \{(y_3 - a_1) + a_1(1 - y_3)\exp(\epsilon_3 t)\}$$
$$\times \{(y_3 - a_1) + (1 - y_3)\exp(t_3 t)\}^{-1}.$$

Hence, if $\beta_3 = 0$, then $\phi_0(t) = \{\phi(t)\}_{(y_3=1)} = \{\phi(t)\}_{(y_3=1, z=0)} = 1$ and $\zeta_{i1}(t) = \{\zeta_i(t)\}_{(y_3=1)}$ satisfies the equation

$$\frac{d}{dt}\zeta_{i1}(t) = b_i\zeta_{i1}^2(t) + [\beta_i z - (b_i + d_i + \beta_i)]\zeta_{i1}(t) + d_i,$$

$$\zeta_{i1}(0) = y_i. \tag{4.5.7}$$

From (4.5.7) and using (3.3.3) of Chapter Three, $\zeta_{i1}(t)$ is given by:

$$\zeta_{i1}(t) = \{a_{2i}(y_i - a_{1i}) + a_{1i}(a_{2i} - y_i)\exp[(a_{2i} - a_{1i})b_i t]\}$$
$$\times \{(y_i - a_{1i}) + (a_{2i} - y_i)\exp[(a_{2i} - a_{1i})b_i t]\}^{-1}, \tag{4.5.8}$$

where $a_{2i} > a_{1i}$ are defined by

$$2b_i a_{2i} = (b_i + d_i + \beta_i - z\beta_i) + h_i(z)$$

and

$$2b_i a_{1i} = (b_i + d_i + \beta_i - z\beta_i) - h_i(z)$$

with

$$h_i(z) = [(b_i + d_i + \beta_i - z\beta_i)^2 - 4b_i d_i]^{1/2}.$$

From the above demonstration, if $\beta_3 \approx 0$ and if $\epsilon_3 t$ is not big for $t_0 \leq t \leq t_1$, then by expanding $\phi_0(t)$ in the Taylor series around $\beta_3 = 0$, $\phi_0(t) = 1 + (\beta_3/\epsilon_3)[1 - \exp(\epsilon_3 t)] + 0(\beta_3^2)$. Hence, if $\beta_3 \approx 0$ and $\omega_i \approx 0$ and if $\epsilon_3 t$ is not big for $t_0 \leq t \leq t_1$, to the order $0(\beta_3 \omega_i)$, $\zeta_{i0}(t)$ satisfies equation (4.5.7) so that $\zeta_{i0}(t)$ is approximately given by formula (4.5.8) for $t_0 \leq t \leq t_1$.

Note that (4.5.8) is in fact the approximation obtained by ignoring the three-step pathways. Since the mutation rates are usually very small, if the cell proliferation rate of $I_3(= I_{12})$ cells is not significantly greater than those of I_i cells, $i = 1, 2$, (4.3.8) would provide a close approximation to $\zeta_{i1}(t)$.

4.5.2 The Age-Specific Incidence Function of Cancer Tumors

To obtain the incidence rate $\lambda(t)$ of a tumor at time t given $N(t_0) = N_0$, we let:

$$\zeta_{i0}(t) = [\zeta_i(t)]_{(y_i = y_3 = 1, z = 0)}, \quad \phi_0(t) = [\phi(t)]_{(y_3 = 1, z = 0)},$$

and

$$\xi_{i,j}(t) = [\partial \zeta_i(t)/\partial y_j]_{(y_i = y_3 = 1, z = 0)}, \quad i = 1, 2, 3, j = 1, 2, 3.$$

From (4.3.2), $\lambda(t)$ is given by:

$$\lambda(t) = N(t)\gamma(t) + \sum_{i=1}^{2} \int_{t_0}^{t} N(u)\alpha_i(u)[\beta_i(u)\xi_{i,i}(t - u)$$

$$+ \beta_3(u)\xi_{i,3}(t - u)]d u, \tag{4.5.9}$$

To compute $\lambda(t)$ by using formula (4.5.9), one would need $\xi_{i,j}(t)$. Now, from equation (4.3.2), by taking derivative with respect to y_j on both sides and setting $y_i = y_3 = 1$ and $z = 0$, one has:

For $i = 1, 2$,

$$\frac{d}{dt} \xi_{i,i}(t) = 2b_i \zeta_{i0}(t)\xi_{i,i}(t) + [\omega_i \phi_0(t) - (b_i + d_i + \omega_i + \beta_i)]\xi_{i,i}(t),$$

$$\xi_{i,i}(0) = 1; \tag{4.5.10}$$

$$\frac{d}{dt} \xi_{i3}(t) = 2b_i \zeta_{i0}(t)\xi_{i,3}(t) + [\omega_i \mu_3(t) + \beta_i]\zeta_{i0}(t) + [\omega_i \phi_0(t)$$

$$- (b_i + d_i + \omega_i + \beta_i)]\xi_{i,3}(t), \xi_{i,3}(0) = 0, \tag{4.5.11}$$

where $\mu_3(t) = [d\phi(t)/d y_3]_{(y_3 = 1, z = 0)}$.

From (4.5.3), one obtains:

$$\frac{d}{dt}\mu_3(t) = 2b_3\phi_0(t)\mu_3(t) - (b_3 + d_3 + \beta_3)\mu_3(t), \quad \mu_3(0) = 1. \quad (4.5.12)$$

From (4.5.12),

$$\mu_3(t) = \exp\left\{\int_0^t [2b_3\phi_0(x) - (b_3 + d_3 + \beta_3)]dx\right\}, \quad (4.5.13)$$

From (4.5.10),

$$\xi_{i,i}(t) = \exp\left\{\int_0^t [2b_i\zeta_{i0}(x) + \omega_i\phi_0(x)]dx - (b_i + d_i + \omega_i + \beta_i)t\right\}, \quad (4.5.14)$$

From (4.5.11)

$$\xi_{i,3}(t) = \int_0^t [\omega_i\mu_3(x) + \beta_i]\zeta_{i0}(x)\xi_{i,i}(x)dx. \quad (4.5.15)$$

To compute $\xi_{i,i}(t)$ and $\xi_{i,3}(t)$, one would need $\phi_0(t)$ and $\zeta_{i0}(t)$. $\phi_0(t)$ is available from (4.5.4) by setting $y_3 = 1$ and $z = 0$. If $\beta_3\omega_i$ is very small and if the cell proliferation rate of $I_3(= I_{12})$ cells is not big, then to order $0(\beta_3\omega_i), i = 1, 2$, an approximation to $\zeta_{i0}(t)$ is available from (4.5.8) by setting $y_i = y_3 = 1$ and $z = 0$. Note further that if $\beta_j \approx 0$ and if ϵ_j is not big for $j = 1, 2, 3$, then $\zeta_{i0}(t) \cong 1$ and $\phi_0(t) \cong 1$. In this case, with $\epsilon_j = b_j - d_j, j = 1, 2, 3$,

$$\xi_{i,i}(t) \cong \exp(\epsilon_i t), \quad i = 1, 2; \quad (4.5.16)$$

$$\mu_3(t) \cong \exp[(\epsilon_3 - \beta_3)t]; \quad (4.5.17)$$

and

$$\begin{aligned}
\xi_{i,3}(t) &\cong \int_0^t \{\omega_i \exp[(\epsilon_3 - \beta_3)x] + \beta_i\} \exp[(\epsilon_i - \beta_i)(t - x)]dx \\
&= [\beta_i/(\epsilon_i - \beta_i)] \{\exp[(\epsilon_i - \beta_i)t] - 1\} \\
&\quad + \{\omega_i/[(\epsilon_3 - \beta_3) - (\epsilon_i - \beta_i)]\} \\
&\quad \times \{\exp[(\epsilon_3 - \beta_3)t] - \exp[(\epsilon_i - \beta_i)t]\}. \quad (4.5.18)
\end{aligned}$$

4.6 A PIECEWISE SEMI-HOMOGENEOUS MODEL FOR THE MODEL IN FIGURE 4.4

In this model, it is assumed that the time interval $[t_0, t]$ is partitioned into k nonoverlapping subintervals $L_j = [t_{j-1}, t_j), j = 1, \ldots, k - 1$ and

$L_k = [t_{k-1}, t_k]$, where $t_k = t$ and that in $L_j, j = 1, \ldots, k, \omega_i(t) = \omega_{ij}, i = 1, 2; b_u(t) = b_{uj}, d_u(t) = d_{uj}$ and $\beta_u(t) = \beta_{uj}, u = 1, 2, 3$. Thus, in essence, this model is the multiple pathway analog of the Tan-Gastardo two-stage model described in Chapter Three. As demonstrated in Chapter Three, this model would be useful in experiments involving initiation and promotion and in studying breast cancer of women.

4.6.1 The Probability Generating Functions (PGF)

To obtain $\psi(t_0, t)$ for this model, let $f_{ij}(s, t) = f_{ij}(y_i, y_3, z; s, t)$ and $\phi_j(y_3, z; s, t) = \phi_j(s, t)$ be the respective solutions of equations (4.6.1) and (4.6.2) given below:

$$\frac{d}{dt} f_{ij}(s, t) = b_{ij} f_{ij}^2(s, t) + [\omega_{ij} \phi_j(s, t) + \beta_{ij} z$$
$$-(b_{ij} + d_{ij} + \omega_{ij} + \beta_{ij})] f_{ij}(s, t) + d_{ij},$$
$$f_{ij}(s, s) = y_i, \quad i = 1, 2; \tag{4.6.1}$$

and

$$\frac{d}{dt} \phi_j(s, t) = b_{3j} \phi_j^2(s, t) + [\beta_{3j} z - (b_{3j} + d_{3j} + \beta_{3j})] \phi_j(s, t)$$
$$+d_{3j}, \quad \phi_j(s, s) = y_3, \quad j = 1, 2, \ldots, k. \tag{4.6.2}$$

(The solutions of (4.6.1) and (4.6.2) always exist; see Moolgavkar and Venzon (34) and Section (4.5). Note also that $\phi_j(s, t) = \phi_j(t - s)$ and $f_{ij}(s, t) = f_{ij}(t - s)$). Put:

$$g_k(s, t) = \phi_k(y_3, z; s, t), h_{i,k}(s, t) = f_{i,k}(y_i, y_3, z; s, t), i = 1, 2;$$

and for $1 \leq j \leq k - 1$ and $t_{j-1} \leq s < t_j$, define :

$$g_j(s, t) = \phi_j[g_{j+1}(t_j, t), z; s, t_j] \tag{4.6.3}$$

and

$$h_{i,j}(s, t) = f_{i,j}[h_{i,j+1}(t_j, t), g_{j+1}(t_j, t), z; s, t_j]. \tag{4.6.4}$$

Then, as shown in Appendix (4A), with $t = t_k$ and $t_{j-1} \leq s < t_j, g_j(s, t)$ and $h_{i,j}(s, t)$ satisfy, respectively the following equations:

$$\frac{\partial}{\partial t} g_j(s, t) = B(k) \frac{\partial}{\partial y_3} g_j(s, t), g_j(s, s) = y_3, \tag{4.6.5}$$

and

$$\frac{\partial}{\partial t} h_{ij}(s,t) = A_i(k) \frac{\partial}{\partial y_i} h_{ij}(s,t) + B(k) \frac{\partial}{\partial y_3} h_{ij}(s,t),$$

$$h_{ij}(s,s) = y_i, \quad i = 1, 2, \tag{4.6.6}$$

where with $t = t_k$,

$$B(k) = (y_3 - 1)[y_3 b_3(t) - d_3(t)] + y_3(z - 1)\beta_3(t)$$
$$= (y_3 - 1)[y_3 b_{3k} - d_{3k}] + y_3(z - 1)\beta_{3k},$$

and

$$A_i(k) = (y_i - 1)[y_i b_i(t) - d_i(t)] + y_i(y_3 - 1)\omega_i(t)$$
$$+ y_i(z - 1)\beta_i(t) = (y_i - 1)[y_i b_{ik} - d_{ik}]$$
$$+ y_i(y_3 - 1)\omega_{ik} + y_i(z - 1)\beta_{ik}.$$

Using (4.6.5) and (4.6.6), $\psi(t_0, t)$ in (4.2.4) then becomes, with $t = t_k$,

$$\phi(t_0, t) = \exp \left\{ (z - 1) \int_{t_0}^{t} N(x)\gamma(x) dx \right.$$

$$+ \sum_{i=1}^{2} \int_{t_0}^{t} N(x)\alpha_i(x)[\zeta_i(x,t) - 1] dx \right\}$$

$$= \exp \left\{ (z - 1) \int_{t_0}^{t} N(x)\gamma(x) dx \right.$$

$$+ \sum_{i=1}^{2} \sum_{j=1}^{k} \int_{t_{j-1}}^{t_j} N(x)\alpha_i(x)[\zeta_i(x,t) - 1] dx \right\}$$

$$= \exp \left\{ (z - 1) \int_{t_0}^{t} N(x)\gamma(x) dx \right.$$

$$+ \sum_{j=1}^{k} \sum_{i=1}^{2} \alpha_{ij} N_{j-1} \int_{t_{j-1}}^{t_j} X(x)[h_{ij}(x,t) - 1] dx \right\}, \tag{4.6.7}$$

where N_{j-1} is the expected number of normal stem cells at t_{j-1} given N_0 normal stem cells at t_0 and $X(x)$ the expected number of normal stem cells at $t = x(x \geq t_{j-1})$ given one normal stem cell at $t = t_{j-1}$.

Formula (4.6.7) is an extension of the PGF of the Tan-Gastardo two-stage model to multiple pathway cases.

Solutions of $g_j(s,t)$ and $h_{ij}(s,t)$

To use formula (4.6.7), one would need $g_j(s,t)$ and $h_{ij}(s,t), j = 1,\ldots,k, i = 1,2$. By definitions of $g_j(s,t)$ and $h_{ij}(s,t)$ as given in formulas (4.6.3) and (4.6.4) respectively, $g_j(s,t)$ and $h_{ij}(s,t)$ will be available if solutions of equations (4.6.1) and (4.6.2) are available.

Let $C_{j2} \geq C_{j1}$ be defined by:

$$2b_{3j}C_{j2} = (b_{3j} + d_{3j} + \beta_{3j} - z\beta_{3j}) + h_{3j}(z)$$

and

$$2b_{3j}C_{j1} = (b_{3j} + d_{3j} + \beta_{3j} - z\beta_{3j}) - h_{3j}(z),$$

where

$$h_{3j}(z) = \{(b_{3j} + d_{3j} + \beta_{3j} - z\beta_{3j})^2 - 4b_{3j}d_{3j}\}^{\frac{1}{2}}.$$

From equation (3.3.1) and using formula (3.3.3) of Chapter Three, the solution $\phi_j(s,t) = \phi_j(y_3,z;s,t)$ of equation (4.6.2) is given by:

$$\phi_j(y_3,z;,s,t) = \{C_{j2}(y_3 - C_{j1}) + C_{j1}(C_{j2} - y_3)\exp[(C_{j2} - C_{j1})b_{3j}t]\}$$
$$\times \{(y_3 - C_{j1}) + (C_{j2} - y_3)\exp[(C_{j2} - C_{j1})b_{3j}t]\}^{-1}, \qquad (4.6.8)$$
$$j = 1,\ldots,k.$$

Using formula (4.6.8), $g_k(s,t) = \phi_k(y_3,z;s,t)$ and $g_j(s,t) = \phi_j[g_{j+1}(t_j,t), z;s,t_j]$ for $j = 1,\ldots,k-1$ are available.

To obtain $h_{ij}(s,t)$, observe that a series solution of equation (4.6.1) is available from Moolgavkar and Venzon (34) but a closed form solution of equation (4.6.1) is very difficult, if not impossible. However, if $d_{ij}[1 - f_{ij}(s,t)] = 0$, then equation (4.6.1) becomes

$$\frac{d}{dt}f_{ij}(s,t) = f_{ij}(s,t)\{b_{ij}f_{ij}(s,t) + \omega_{ij}\phi_j(s,t)$$
$$+ \beta_{ij}z - (b_{ij} + \omega_{ij} + \beta_{ij})\}, f_{ij}(s,s) = y_i, i = 1,2, j = 1,\ldots,k. \qquad (4.6.9)$$

The solution of equation (4.6.9) is given by:

$$f_{ij}(s,t) = f_{ij}(y_i,y_3,z;s,t) = y_i\theta_{ij}(s,t)\{1 - y_iU_{ij}(s,t)\}^{-1}, \qquad (4.6.10)$$

where

$$\theta_{ij}(s,t) = \exp\{t\beta_{ij}(z-1) - tb_{ij} + \omega_{ij}\int_s^t [\phi_j(s,x) - 1]dx\}$$

and

$$U_{ij}(s,t) = b_{ij} \int_s^t \theta_{ij}(s,x)dx.$$

(For details, see Section (4.5.1)).

Thus, if $d_{ij}[1 - f_{ij}(s,t)] \approx 0$, close approximations to $h_{i,k}(s,t) = f_{ik}(y_i, y_3, z; s, t)$ and

$$h_{ij}(s,t) = f_{ij}[h_{i,j+1}(t_j,t), g_{j+1}(t_j,t), z; s, t_j], j = 1,\ldots, k - 1, i = 1, 2$$

can readily be derived by using formula (4.6.10).

4.6.2 The Age-Specific Incidence Functions of Tumors

By (4.3.2), with $t = t_m$,

$$\lambda(t) = N(t)\gamma(t) + \sum_{i=1}^2 \int_{t_0}^t N(x)\alpha_i(x)[\beta_i(t)u_{i,i}(x,t) + \beta_3(t)u_{i,3}(x,t)]dx$$

$$= N(t)\gamma(t) + \sum_{i=1}^2 \sum_{j=1}^k \int_{t_{j-1}}^{t_j} N(x)\alpha_i(x)[\beta_i(t)u_{i,i}(x,t)$$

$$+ \beta_3(t)u_{i,3}(x,t)]dx = N(t)\gamma(t) + \sum_{i=1}^2 \sum_{j=1}^k \alpha_{ij}N_{j-1}$$

$$\times \int_{t_{j-1}}^{t_j} X_{j-1}(x)[\beta_{ik}u_{i,i(j)}(x,t) + \beta_{3k}u_{i,3(j)}(x,t)]dx, \tag{4.6.11}$$

where for $t_{j-1} \le x < t \le t_j, j = 1,\ldots, k$,

$$u_{i,u(j)}(x,t) = [\partial \zeta_i(x,t)/\partial y_u]_{(y_i = y_3 = 1, z = 0)}$$
$$= [\partial h_{i,j}(x,t)/\partial y_u]_{(y_i = y_3 = 1, z = 0)}, u = i \text{ or } u = 3.$$

To compute $\lambda(t)$ by (4.6.11), one would need the solution of $h_{ij}(x,t)$. Since the solution of equation (4.6.1) is very complicated, we thus use the approximation (4.3.7). This approximation is expected to be very close if $\beta_i(t)$ is very small and if $\epsilon_i(t)$ is not big for all $t \ge t_0$.

Now, by (4.3.7), with $t = t_k$,

$$\lambda(t) \approx N(t)\gamma(t) + \sum_{i=1}^2 \int_{t_0}^t N(x)\alpha_i(x)[\beta_i(t)m_{i,i}(x,t)$$
$$+ \beta_3(t)m_{i,3}(x,t)]dx$$

$$= N(t)\gamma(t) + \sum_{i=1}^{2}\sum_{j=1}^{k} \alpha_{ij}N_{j-1}\int_{t_{j-1}}^{t_j} X_{j-1}(x)[\beta_{ik}m_{i,i(j)}(x,t)$$

$$+ \beta_{3k}m_{i,3(j)}(x,t)]dx, \qquad (4.6.12)$$

where for $t_{j-1} \le x < t_j, j = 1,\ldots,k,$

$$m_{i,i}(x,t) = m_{i,i(j)}(x,t) = \exp\{\int_x^t \epsilon_i(y)dy\}$$

and

$$m_{i,3}(x,t) = m_{i,3(j)}(x,t) = \int_x^t \omega_i(y)m_{i,i}(x,y)\exp[\int_y^t \epsilon_3(u)du]dy.$$

Note that (4.6.12) is obtained from (4.6.11) by approximating $u_{i,u(j)}(x,t)$ by $m_{i,u(j)}(s,t)$ for $u = i$ and $u = 3$ respectively.

To compute $\lambda(t)$ by using formular (4.6.12), observe that for $t_{j-1} \le x < t_j, j = 1,\ldots,k,$ one has:

$$m_{i,i(j)}(x,t) = \exp\{\int_x^t \epsilon_i(y)dy\} = \exp\{\int_x^{t_j} \epsilon_i(y)dy$$

$$+ \sum_{u=j+1}^{k}\int_{t_{u-1}}^{t_u} \epsilon_i(y)dy\} = \exp\{\epsilon_{ij}(t_j - x) + \sum_{u=j+1}^{k} \epsilon_{iu}\tau_u\},$$

where $\tau_u = t_u - t_{u-1};$

$$m_{i,3(j)}(x,t) = \int_x^t \omega_i(y)m_{i,i(j)}(x,y)\exp[\int_y^t \epsilon_3(u)du]dy$$

$$= \{\int_x^{t_j} + \sum_{u=j+1}^{k}\int_{t_{u-1}}^{t_u}\}\omega_i(y)m_{i,i(j)}(x,y)\exp[\int_y^t \epsilon_3(u)du]dy$$

$$= \omega_{ij}\int_x^{t_j} \exp[\epsilon_{ij}(y-x) + \epsilon_{3j}(t_j - y) + \sum_{v=j+1}^{k} \epsilon_{3v}\tau_v]dy$$

$$+ \sum_{u=j+1}^{k} \omega_{iu}\int_{t_{u-1}}^{t_u} \exp[\epsilon_{ij}(t_j - x) + \sum_{v=j+1}^{u-1} \epsilon_{iv}\tau_v$$

$$+ \epsilon_{iu}(y - t_{u-1}) + \epsilon_{3u}(t_u - y) + \sum_{v=u+1}^{k} \epsilon_{3v}\tau_v]dy$$

$$= \omega_{ij} \exp\left(\sum_{v=j+1}^{k} \epsilon_{3u}\tau_v\right)(\epsilon_{ij} - \epsilon_{3j})^{-1}\{\exp[\epsilon_{ij}(t_j - x)]$$

$$- \exp[\epsilon_{3j}(t_j - x)]\} + \sum_{u=j+1}^{k} \omega_{iu} \exp[\epsilon_{ij}(t_j - x)]$$

$$+ \sum_{v=j+1}^{u-1} \epsilon_{iv}\tau_v + \sum_{v=u+1}^{k} \epsilon_{3v}\tau_v](\epsilon_{iu} - \epsilon_{3u})^{-1}$$

$$\times [\exp(\epsilon_{iu}\tau_u) - \exp(\epsilon_{3u}\tau_u)]. \tag{4.6.13}$$

4.7 A MULTIPLE PATHWAY MODEL INVOLVING A ONE-STAGE MODEL AND A TWO-STAGE MODEL

Suppose that there is only one two-stage pathway in the model of Figure 4.4. Then, there is only one type of intermediate cell which we shall refer to as I cells. In this case, the model is considerably simplified and many results of Chapter Three are directly applicable.

To illustrate, we let the birth rate and the death rate of the I cells be $b(t)$ and $d(t)$ respectively and denote by $\bar{\alpha}(t)$ the mutation rate of the first event and $\bar{\beta}(t)$ the mutation rate of the second event in the two-stage pathway. Further, we assume conditions (1), (5) and (6) of Section (4.2.1) and replace conditions (2), (3) and (4) by the following three conditions:

(b)' The I cells follow a nonhomogeneous Feller-Arley birth-death process for cell proliferation and cell differentiation with birth rate $b(t)$ and death rate $d(t)$.

(c)' Given the one-step pathway from normal stem cells to tumor cells, the probability that a normal stem cell at time t yields one normal stem cell and one tumor cell at time $t + \Delta t$ is $\bar{\gamma}(t)\Delta t + o(\Delta t)$. Given the two-step pathway, the probability that a normal stem cell at time t yields one normal stem cell and one I cell at time $t + \Delta t$ is $\bar{\alpha}(t)\Delta t + o(\Delta t)$. In what follows, we denote by $\gamma(t) = p(t)\bar{\gamma}(t)$ and $\alpha(t) = [1 - p(t)]\bar{\alpha}(t)$, where $p(t)$ is the probability for the one-step pathway at time t.

(d)' The probability that an I cell at time t yields one I cell and one tumor cell at time $t + \Delta t$ is $\beta(t)\Delta t + o(\Delta t)$. Note that $\beta(t) = p_1(t)\bar{\gamma}(t) + [1 - p_1(t)]\bar{\beta}(t)$, where $p_1(t)$ is the probability for the one-step pathway. ($p_1(t)$ may equal $p(t)$).

Given these assumptions, one may immediately derive results from previous Sections. In what follows, we let $N(t), I(t)$ and $T(t)$ denote the numbers of normal stem cells, I cells and tumors at time t, respectively.

4.7.1 The Probability Generating Functions

Let $\psi(y,z;t_0,t) = \psi(t_0,t)$ be the PGF of $I(t)$ and $T(t)$ given $N(t_0) = N_0$ and $\phi(y,z;t_0,t) = \phi(t_0,t)$ the PGF of $I(t)$ and $T(t)$ given one I cell at time $t_0(t \geq t_0)$. Then, from Theorems (4.2.2) and (4.2.1), we have:

THEOREM (4.7.1) Suppose that $N(t)\gamma(t)$ and $N(t)\alpha(t)$ are finite for all $t \geq t_0$ and that during $[t, t + \Delta t]$, the numbers of mutations from normal stem cells to tumor cells and from normal stem cells to I cells follow independent Poisson distributions with parameters $N(t)\gamma(t)\Delta t + o(\Delta t)$ and $N(t)\alpha(t)\Delta t + o(\Delta t)$ respectively. Then, under conditions (1), (b)', (c)', (d)', (5) and (6), $\psi(t_0,t)$ is given by:

$$\psi(t_0,t) \cong \exp \{(z-1) \int_{t_0}^{t} N(x)\gamma(x)dx$$

$$+ \int_{t_0}^{t} N(x)\alpha(x)[\phi(x,t)-1]dx\} \quad (4.7.1)$$

where $N(x)$ is the expected number of normal stem cells at time t given $N(t_0) = N_0$.

The proof of Theorem (4.7.1) is exactly the same as that of Theorem (4.2.2).

THEOREM (4.7.2) Under conditions (b)', (c)', (d)', (5) and (6), $\phi(t_0,t)$ satisfies the following first order partial differential equations:

$$\frac{\partial}{\partial t}\phi(t_0,t) = \{(y-1)[yb(t) - d(t)] + y(z-1)\beta(t)\}$$

$$\times \frac{\partial}{\partial y}\phi(t_0,t), \phi(t_0,t_0) = y. \quad (4.7.2)$$

The proof of Theorem (4.7.2) is exactly the same as that of Theorem (4.2.1); see also Section (3.2.2) of Chapter Three. Note that (4.7.2) is exactly the same equation as equation (3.2.3) of Chapter Three.

From (4.7.1) and (4.7.2), obviously the results of Chapter Three can be applied in a very straightforward manner to obtain incidence functions of tumors and the probability distributions of the number of tumors. In the next section, Section (4.7.2), we give the incidence function of tumors.

4.7.2 The Age-Specific Incidence Functions of Tumors

Let $\lambda(t)$ be the incidence rate of tumors at time t given $N(t_0) = N_0$.

From the results of Section (4.3) and the results of Section (3.4) of Chapter Three,

$$\lambda(t) = N(t)\gamma(t) + \beta(t) \int_{t_0}^{t} N(x)\alpha(x)\mu_2(x,t)dx$$

where

$$\mu_2(x,t) = [\partial\phi(x,t)/\partial y]_{(y=1,z=0)}.$$

If $\beta(t)$ is very small and if $\epsilon(t) = b(t) - d(t)$ is not big for all $t \geq t_0$, then as shown in Section (3.4.1) of Chapter Three, a close approximation to $\lambda(t)$ is

$$\lambda(t) \cong N(t)\gamma(t) + \beta(t) \int_{t_0}^{t} N(x)\alpha(x) \exp\left\{ \int_{x}^{t} [b(v)-d(v)]dv\right\}dx, \quad (4.7.3)$$

4.7.3 Some Special Cases

The Homogeneous Model

If $\alpha(t) = \alpha, \beta(t) = \beta, b(t) = b$ and $d(t) = d$ are independent of t, then the two-stage pathway becomes the MVK two-stage model.

In this case, $\phi(t_0,t) = \phi(t - t_0); \phi(t_0,t)$ also satisfies the following Ricatti equation:

$$\frac{d}{dt}\phi(t_0,t) = b\phi^2(t_0,t) + [\beta z - (b+d+\beta)]\phi(t_0,t) + d, \; \phi(t_0,t_0) = y. \quad (4.7.4)$$

Taking derivative with respect to y on both sides of equation (4.7.4) and setting $y = 1$ and $z = 0$, we obtain:

$$\frac{d}{dt}\mu_2(x,t) = 2b\phi_{10}(s,t)\mu_2(x,t) - [b+d+\beta]\mu_2(x,t), \; \mu_2(x,x) = 1,$$

where

$$\phi_{10}(x,t) = [\phi(x,t)]_{(y=1,z=0)}.$$

It follows that

$$\mu_2(x,t) = \exp\{-(b+d+\beta)(t-x) + 2b\int_x^t \phi_{10}(x,u)du\}$$

If $\beta \approx 0$ and if $\epsilon = b - d$ is not big so that $\phi_{10}(x,u) \approx 1$, then

$$\mu_2(x,t) \approx \exp\{(b-d-\beta)(t-x)] \approx \exp[(b-d)(t-x)].$$

Thus, if $\beta \approx 0$ and if $\epsilon = b - d$ is not big,

$$\lambda(t) \approx N(t)\gamma(t) + \alpha\beta\int_{t_0}^t N(x)\exp[(b-d)(t-x)]dx.$$

The Case $d(t) = 0$

As shown in Chapter Three, much evidence exists indicating that the normal stem cells become immortalized by the loss of differentiation capability (5, 25, 37). These results suggest that one may perhaps assume $d(t) \approx 0$ for all $t \geq t_0$. In this case, $\phi(t_0,t)$ is given by:

$$\phi(t_0,t) = y/\{\xi(t_0,t) - yU(t_0,t)\},$$

where

$$\xi(t_0,t) = \exp\{\int_{t_0}^t [b(x) + \beta(x)]dx\}$$

and

$$U(t_0,t) = \int_{t_0}^t b(x)\xi(t_0,x)dx.$$

Hence $\lambda(t)$ reduces to:

$$\lambda(t) = N(t)\gamma(t) + \beta(t)\int_{t_0}^t N(x)\alpha(x)\xi(x,t)[\xi(x,t) - U(x,t)]^{-2}dx.$$

If $\beta(t) \approx 0$ and if $b(t)$ is not big for all $t \geq t_0$, then

$$\xi(x,t) = \exp[\int_x^t b(u)du] \quad \text{and} \quad U(x,t) = \int_x^t b(u)\xi(x,u)du$$

$$= \int_x^t d\exp[\int_x^u b(v)dv] = \exp[\int_x^t b(v)dv] - 1.$$

It follows that if $\beta(t) \approx 0$ and if $b(t)$ is not big for all $t \geq t_0$,

$$\lambda(t) \approx N(t)\gamma(t) + \beta(t)\int_{t_0}^t N(x)\alpha(x)\exp[\int_x^t b(v)dv]dx. \qquad (4.7.5)$$

The Piece-wise Homogeneous Model

In this model, the time interval $[t_0, t]$ is partitioned into k nonoverlapping intervals $L_j = [t_{j-1}, t_j), j = 1, \ldots, k-1, L_k = [t_{k-1}, t_k], t = t_k$, such that in $L_j, \alpha(t) = \alpha_j, \beta(t) = \beta_j, b(t) = b_j, d(t) = d_j, j = 1, \ldots, k.$

Let $y_{2j} > y_{1j}$ be defined by $2b_j y_{2j} = (b_j + d_j + \beta_j - z\beta_j) + h_j(z)$ and $2b_j y_{1j} = (b_j + d_j + \beta_j - z\beta_j) - h_j(z)$, where $h_j(z) = [(b_j + d_j + \beta_j - z\beta_j)^2 - 4b_j d_j]^{1/2}$ and put

$$\phi_j(y,z;s,t) = \phi(y,z;t-s)$$
$$= \{y_{2j}(y - y_{1j}) + y_{1j}(y_{2j} - y) \exp [(y_{2j} - y_{1j})b_j(t-s)]\}$$
$$\times \{(y - y_{1j}) + (y_{2j} - y) \exp [(y_{2j} - y_{1j})b_j(t-s)]\}^{-1}.$$

Let $g_k(s,t) = \phi_k(y,z;t-s)$, $t_{k-1} \le s < t$, and for $t_{j-1} \le s < t_j, j = 1, \ldots, k-1,$

$$g_j(s,t) = \phi_j[g_{j+1}(t_j,t),z;s,t_j].$$

Then, as shown in Chapter Three,

$$\psi(t_0,t) \cong \exp \{ \int_{t_0}^{t} N(x)\gamma(x)dx$$

$$+ \sum_{i=1}^{k} \int_{t_{i-1}}^{t_i} N(x)\alpha(x)[g_i(x,t) - 1]dx\}. \quad (4.7.6)$$

Let $g_{j0}(s,t) = [g_j(s,t)]_{(y=1,z=0)}$. Then, as shown in Chapter Three,

$$\lambda(t) = N(t)\gamma(t) + \beta(t) \sum_{j=1}^{k} \int_{t_{j-1}}^{t_j} N(x)\alpha(x)m_j(t_j - x)dx$$

$$\times \left[\prod_{u=j+1}^{k} m_u(\tau_u) \right], \quad (4.7.7)$$

where $\tau_u = t_u - t_{u-1}$, and

$$m_v(x) = \exp\{-(b_v + d_v + \beta_v)x + 2b_v \int_{0}^{x} g_{v0}(0,u)du\}, \quad v = 1, \ldots, k.$$

If $\beta_v \approx 0$ and if $\epsilon_v = b_v - d_v$ is not big for all $v = 1, \ldots, k$, then $g_{v0}(0,u) \approx 1$ and hence $m_v(x) \approx \exp[(b_v - d_v)x]$.

The second term in (4.7.7) was first obtained by Tan and Gastardo (46). Note that the above models are useful in experiments involving initiators and promoters.

4.8 SOME POSSIBLE APPLICATIONS

Since multiple pathway models of carcinogenesis have just been proposed, attempts to apply these models to solve many practical problems have not yet been made and procedures to apply these models to many practical problems remain to be developed. However, it is appropriate to point out that multiple pathway models are useful in the following contexts which may open up new research frontiers for carcinogenesis models.

Interpretation of Inconsistent Cancer Data

The existence of much inconsistent data in some human cancers suggests that for these cancers multiple pathway models prevail. For instance, for the development of nodular melanoma in skin cancer, some data seemed to suggest sunlight as causal agents (11, 13, 18, 21, 41, 43) while other data tended to exclude sunlight as a major cause of nodular melanoma (15, 26, 27). To interpret these inconsistent data, Holman et al. (16) observed that nodular melanoma arose either from a pathway involving HMF (Hutchinson's Melanotic Freckle) or from a pathway involving SSM (Superficial Spreading Melanoma) and UCM (Unclassified Type Melanoma). It appears that HMF or other skin cancers than melanomas occur as functions of accumulative DNA damages for which ultraviolet (UV) radiation is a causing agent; but in the pathway involving SSM and UCM, UV radiation functions as a promoter for initiated nevus cells which are mutated cells arising from melanocytes for which ultraviolet radiation may also serve as an initiator. These observations together with the two pathway model for melanoma development with one involving HMF and the other involving SSM and UCM (non-HMF) would explain the observed pattern of incidence curves of melanoma at different sites (15, 27). Thus, for the skin of the head and neck where HMF is predominant, the incidence curve of melanomas increases with age and/or continuous exposure to sunlight, whereas at other body sites than the head and neck, where HMF is rare (SSM and UCM are predominant), the incidence curve of melanomas rises steeply in the early adult years, reaches a plateau in middle life and then falls. These results are consistent with the observations made by Holman, Mulroney and Armstrong (15) and by Magrus (27). Also, since the occurrence of tanning would reduce the amount of UV radiation reaching the melanocytes, it is anticipated that intermittent (recreational) stimulation of the melanocyte system by UV radiation may have a more powerful promotional effect than continuous stimulation on initiated nevus cells. As a net effect, continuous exposure to sunlight leading to persistent tanning would yield a greater protective effect against non-HMF than HMF melanoma while increased intermittent (recreational) exposure to sunlight would cause an increase in the proportion of non-HMF melanomas. This is consistent with the observations made by Magrus (27) that increased intermittent exposure to sunlight yielded greater proportional increases with time in melanomas of the trunk and lower limb compared with the melanomas of the head and neck. This would also help explain the higher incidence of melanoma in urban populations than in rural populations (26) and in indoor rather than outdoor workers (15) and the observations that melanomas arises most commonly on the trunk in man and lower limbs in women and affects a younger group of patients (15).

As another example, consider the breast tumors of women. Conflicting results have been reported for oncogene expressions in human breast tumors. For example, Clair et al (8) reported a positive correlation between the expression of p21 of the ras oncogene and the progression of breast cancer; yet, Chesa et al (7) reported a lack of association between p21 ras expression and proliferation and malignancy. Horan-Hand et al (17) noted an elevated level of p21 of ras in breast cancers compared with normal tissues, but increased levels were also detected in dysplastic lesions. Noting the homogeneity of the neoplastic population and the very high incidence of breast cancers in women, Medina (30) suggested multiple pathway mechanisms for neoplastic development of human breast cancers, with one pathway involving ras while other pathways do not involve ras.

Risk Assessments of Environmental Agents

By comparing results from the one-hit model, multi-hit model, Weibull model and classical Armitage–Doll multistage model, Van Ryzin (52, 53) has shown that for risk assessment of carcinogens, different models give very different results. It follows that for obtaining correct risk assessment, the adopted model should truly reflect the biological mechanisms. In this sense then, the multiple pathway models should be useful for risk assessment of environmental agents. Thus, it is desirable to incorporate the multiple pathway models into risk assessment procedures for assessing the risk of environmental agents. Since multiple pathway models are just being proposed, no attempt has been made to use multiple pathway models to assess risks of environmental agents. In this respect, much new research is desired.

Assessment of Robustness and Efficiencies of Risk Assessment Procedures

As another application, one may use the multiple pathway models as a basis to assess robustness and efficiencies of risk assessment procedures. In fact, by assuming some multiple pathway models one can readily generate Monte Carlo data by using a computer; by using these computer generated data one may then evaluate the efficiencies and robustness of many risk assessment procedures for assessing the risk of environmental agents. Recently, attempts have been made to incorporate the Moolgavkar-Venzon-Knudson two-stage models into risk assessment procedures for assessing the risk of environmental agents (Portier and Bailer (38), Moolgavkar and Luebeck (36), Tan and Starr (49)). By adopting the multiple pathway models, one may evaluate the robustness and the efficiencies of the procedures developed by Moolgavkar and Luebeck (36).

4.9 CONCLUSIONS AND SUMMARY

Much biological evidence exists indicating that in many cases cancer tumors develop from multiple pathways rather than from a single pathway. In this chapter we thus proceed to develop the mathematical theories of multiple pathway models of carcinogenesis. Since mutation rates are usually very small so that pathways involving three or more stages are usually dominated by one-stage and two-stage pathways, in the mathematical derivations we thus restrict ourselves only to multiple pathway models involving one-stage models and two-stage models.

To start the chapter, in Section (4.1) we present some biological evidence supporting multiple pathway models involving one-stage models and two-stage models. As illustrated in Section (4.1), the one-stage models arise either through the action of dual oncogenes such as the SV40 large T antigen oncogene or through the elimination of the inhibitory effects of neighboring normal stem cells whereas the multiple pathway models involving two-stage models arise because oncogenes in the same class share some common properties and functions. Based on the biological evidence, we thus propose in Section (4.2) a stochastic multiple pathway model involving one-stage and two two-stage models. Under some general conditions, we derive in Section (4.2) some differential equations for the probability generating functions (PGF) of normal stem cells, intermediate cells and cancer tumors; the PGF of intermediate cells and cancer tumors is derived under some general conditions and is given in formula (4.2.7). The results of Section (4.2) show that under some general conditions the one-stage pathway is described by a non-homogeneous Poisson process whereas the two-stage pathways are described by nonhomogeneous filtered Poisson processes.

By using formula (4.2.7), we derive in Section (4.3) the age-specific incidence function of tumors and give an approximation to this incidence function by assuming that the mutation rates are very small and the cell proliferation rates are not big (≤ 0.1). Again by using formula (4.2.7), we derive in Section (4.4) the expected number of tumors and the variances and covariances of the number of tumors.

Having derived results for the general model given in Figure 4.4, in Sections (4.5)–(4.7) we then consider some special cases in each of which we derive the PGF and the incidence functions of tumors. These special cases include a semi-homogeneous model, a piece-wise semi-homogeneous model and a multiple pathway model involving one one-stage and one two-stage models. Finally in Section (4.8), some possible applications of the multiple pathway models are indicated.

APPENDIX

Proof of (4.6.5) and (4.6.6). Given the model in Section (4.6), we now show by mathematical induction that for $t = t_k$ and $t_{j-1} \leq s < t_j, j = 1,\ldots,k$, the functions $g_j(s,t)$ in (4.6.3) and $h_{ij}(s,t)$ in (4.6.4) satisfy, respectively, the following partial differential equations:

$$\frac{\partial}{\partial t} g_j(s,t) = B(k) \frac{\partial}{\partial y_3} g_j(s,t), \; g_j(s,s) = y_3, \qquad (4.A.1)$$

and

$$\frac{\partial}{\partial t} h_{ij}(s,t) = A_i(k) \frac{\partial}{\partial y_i} h_{ij}(s,t) + B(k) \frac{\partial}{\partial y_3} h_{ij}(s,t),$$

$$h_{ij}(s,s) = y_i, \quad (4.A.2)$$

where $A_i(k)$ and $B(k)$ are given in (4.6.5) and (4.6.6).

Proof by Mathematical Induction. By (4.2.4) and (4.2.5) of Theorem (4.2.1), $g_j(s,t)$ and $h_{i,j}(s,t)$ satisfy, respectively (4.A.1) and (4.A.2) if $j = k$.

Suppose that $g_j(s,t)$ and $h_{i,j}(s,t)$ satisfy, respectively, (4.A.1) and (4.A.2) for $j = m$, where $1 < m \le k$. It suffices then to show that $g_j(s,t)$ and $h_{i,j}(s,t)$ satisfy, respectively, (4.A.1) and (4.A.2) for $j = m - 1$.

To show (4.A.1) and (4.A.2) for $j = m - 1$ given the results for $j = m$, observe first that by definition (4.6.3) and (4.6.4),

$$g_{m-1}(s,t) = \phi_{m-1}[g_m(t_{m-1},t),z;t_{m-1} - s]$$

and

$$h_{i,m-1}(s,t) = f_{i,m-1}[f_{i,m}(t_{m-1},t),g_m(t_{m-1},t),z;t_{m-1} - s].$$

Hence, by the chain rule,

$$\frac{\partial}{\partial t} g_{m-1}(s,t) = \{\partial g_{m-1}(s,t)/\partial g_m(t_{m-1},t)\} \frac{\partial}{\partial t} g_m(t_{m-1},t)$$

$$= B(k)\{\partial g_{m-1}(s,t)/\partial g_m(t_{m-1},t)\} \frac{\partial}{\partial y_3} g_m(t_{m-1},t)$$

$$= B(k) \frac{\partial}{\partial y_3} g_{m-1}(s,t), \quad g_{m-1}(s,s) = y_3. \tag{4.A.3}$$

Similarly, by the chain rule and noting that $g_m(t_{m-1},t)$ is independent of y_i,

$$\frac{\partial}{\partial t} h_{i,m-1}(s,t) = \{\partial h_{i,m-1}(s,t)/\partial h_{i,m}(t_{m-1},t)\} \frac{\partial}{\partial t} h_{i,m}(t_{m-1},t)$$

$$+ \{\partial h_{i,m-1}(s,t)/\partial g_m(t_{m-1},t)\} \frac{\partial}{\partial t} g_m(t_{m-1},t)$$

$$= \{\partial h_{i,m-1}(s,t)/\partial h_{i,m}(t_{m-1},t)\}\{A_i(k) \frac{\partial}{\partial y_i} h_{i,m}(t_{m-1},t)$$

$$+ B(k) \frac{\partial}{\partial y_3} h_{i,m}(t_{m-1},t)\}$$

$$+ \{\partial h_{i,m-1}(s,t)/\partial g_m(t_{m-1},t)\} B(k) \frac{\partial}{\partial y_3} g_m(t_{m-1},t)$$

$$= A_i(k)\{\partial h_{i,m-1}(s,t)/\partial h_{i,m}(t_{m-1}t\} \frac{\partial}{\partial y_i} h_{i,m}(t_{m-1},t)$$

$$+ B(k)\{[\partial h_{i,m-1}(s,t)/\partial h_{i,m}(t_{m-1},t)] \frac{\partial}{\partial y_3} h_{i,m}(t_{m-1},t)$$

$$+[\partial h_{i,m-1}(s,t)/\partial g_m(t_{m-1},t)]\frac{\partial}{\partial y_3}g_m(t_{m-1},t)\} \qquad (4.A.4)$$

$$= A_i(k)\frac{\partial}{\partial y_i}h_{i,m-1}(s,t) + B(k)\frac{\partial}{\partial y_3}h_{i,m-1}(s,t),\ h_{i,m-1}(s,s) = y_i.$$

From (4.A.3) and (4.A.4), $g_j(s,t)$ and $h_{i,j}(s,t)$ also satisfy, respectively (4.A.1) and (4.A.2) for $j = m - 1$ if the results hold for $j = m$.

REFERENCES

1. Adkins, B., Leutz, A., and Graf, T., Autocrine growth induced by src-related oncogenes in transformed chicken myeloid cells, Cell **39** (1984), 439–445.
2. Albert, M. E., "The time to tumor approach. In: *Mechansisms of DNA Damage and Repair: Implications For Carcinogenesis And Risk Assessment*," Edited by M.G. Simic, L. Grossman and A. C. Upton, Plenum Press, N. Y. (1986).
3. Bechade, C., et al., Induction of proliferation or transformation of neuroretina cells by the mil viral oncogenes, Nature **316** (1985), 559–562.
4. Bignami, M. , et al., Differential influence of adjacent normal cells on the proliferation of mammalian cells transformed by the viral oncogenes myc, ras and src, Oncogene **2** (1988), 509–514.
5. Buick, R. N. and Pollak, M. N., Perspective on clonogenic tumor cells, stem cells and oncogenes, Cancer Res **44** (1984), 4909–4918.
6. Butel, J. S., SV40 large T antigen: Dual oncogene, Cancer Survey **5** (1986), 343–366.
7. Chesa, P. G. et al., Expression of $p21^{ras}$ in normal and malignant human tissues: Lack of association with proliferation and malignancy, Proc. Natl. Acad. Sci. U.S.A **84** (1987), 3234–3238.
8. Clair, T., Miller, W. R. and Cho-chung, Y. S., Prognostic significance of the expression of a ras protein with a molecular weight of 21,000 by human breast cancer, Cancer Res. (1987), 5290–5293.
9. Compere, S. J. et al., The ras and myc oncogenes cooperate in tumor induction in many tissues when introduced into midgestation mouse embryos by retroviral vectors, Proc. Natl. Acad. Sci. USA **86** (1989), 2224–2228.

10. Detto, G. P. , Weinberg, R. A. and Ariza, A., Malignant transformation of mouse primary keratinocytes by HaSV and its modulation by surrounding normal cells, Proc. Natl. Acad. Sci. **85** (1988), 6389–6393.

11. Elwood, J. M., et al., Relationship of melanoma and other skin cancer mortality to latitude and ultraviolet radiation in the United States and Canada, Int. J. Epidemiol. **3** (1974), 325–332.

12. Foulds, L., "Neoplastic Development Vol 2," Academic Press, N. Y., 1975.

13. Gellin, G. A., Kopf, A. W. and Garfinkel, L., Malignant melanoma. A controlled study of possible associated factors, Arch. Dermatol. **99** (1969), 43–48.

14. Glaichenhaus, N., et al., Cooperation between multiple oncogenes in rodent embryo fibroblasts: An experimental model of tumor progression, Advances in Cancer Research **45** (1985), 291–305.

15. Holman, C. D., Mulroney, C. D. and Armstrong, B. K., Epidemiology of preinvasive and invasive malignant melanoma in Western Australia, Int. J. Cancer **25** (1980), 317–323.

16. Holman, L. D'Arcy J., Armstrong, B. K. and Heenan, P. J., A theory of the etiology and pathologenesis of human cutaneous malignant melanoma, Jour. Nat. Cancer Inst. **71** (1983), 651–656.

17. Horan-Hand, P. et al., Quantitation of Harvey ras p21 enhanced expression in human breast and colon carcinomas, Jour. Natl. Cancer Inst. **79** (1987), 59–65.

18. Houghton, A., Munster, E. W. and Viola, M. V., Increased incidence of malignant melanoma after peaks of sunspot activity, Lancet **1** (1978), 759–760.

19. Keath, E. J., Caimi, P. G. and Cole, M.D., Fibroblast lines expressing activated c-myc oncogenes are tumorigenic in nude mice and syngeneic animals, Cells **39** (1984), 339–348.

20. Knudson, A. G., Hereditary cancer, oncogenes and antioncogenes, Cancer Res **45** (1985), 1437–1443.

21. Lancaster, H. O. and Nelson, J., Sunlight as a cause of melanoma: A clinical survey, Med. J. Aust. **1** (1957), 452–456.

22. Land, H., Parada, L. F. and Weinberg, R. A., Tumorigenic conversion of primary embryo fibroblasts requires at least two cooperating oncogenes, Nature **304** (1983), 596–601.

23. Land, H. , et al., Behavior of myc and ras oncogenes in transformation of rat embryo fibroblasts, Mol. Cell Biol. **6** (1986), 1917–1925.

24. Landord, R. E., Wong, O. and Butel, J. S., Differential ability of a T-antigen transport-defective mutant of simiar virus 40 to transform primary and established rodent cells, Molecular Cellular Biology **5** (1985), 1043–1050.

25. Mackillop, W. J., Ciampi, A. and Buck, R. N., A stem cell model of human tumor growth: Implications for tumor cell clonogenic assays, J. Nat. Cancer Inst. **70** (1983), 9–16.

26. Magnus, K., Incidence of malignant melanoma of the skin in Norway 1955-1977, Cancer **32** (1973), 1275–1286.

27. Magnus, K., Habits of sun exposure and risk of malignant melanoma: An analysis of incidence rates in Norway 1955-1977 by cohort, sex, and age and primary tumor site, Cancer **48** (1981), 2329–2335.

28. Marshall, C. J. and Ridby, P. W. J. C., Viral and cellular genes involved in oncogenesis, Cancer Survey **3** (1984), 183–214.

29. Marshal, C. J., Oncogenes, J. Cell Sci. Suppl. **4** (1986), 417–430.

30. Medina, D., The preneoplastic state in mouse mammary tumorigenesis, Carcinogenesis **9** (1988), 1113–1119.

31. Mishima, Y., Melanocytic and nevocytic malignant melanoma. Cellular and subcellular differentiation, Cancer **20** (1967), 632–649.

32. Moolgavkar, S. H., Carcinogenesis modeling: From molecular biology to epidemiology, Ann. Rev. Public Health **7** (1986), 151–169.

33. Moolgavkar, S. H., Hormones and multistage carcinogenesis, Cancer Survey **5** (1986), 635–648.

34. Moolgavkar, S. H. and Venzon, D. J., Two event models for carcinogenesis: Incidence curves for childhood and adult cancer, Math Biosciences **47** (1979), 55–77.

35. Moolgavkar, S. H. and Knudson, A. G., Mutation and cancer: A model for human carcinogenesis, Jour. Nat. Cancer Inst. **66** (1981), 1037–1052.

36. Moolgavkar, S. H. and Luebeck, E. G., "A biologically motivated parametric model for the analysis of animal tumorigenicity data," Fred Hutchinson Cancer Research Center, University of Washington, Seattle, WA, 1989.

37. Oberley, L. W. and Oberley, T. D., The role of superoxide dismutase and gene amplification in carcinogenesis, J. Theor. Biology **106** (1984), 403-422.

38. Portier, C.J. and Bailer, A.J., "Two-stage models of tumor incidence for historical control animals in the national toxicology program's

carcinogenicity experiments," National Institute of Environmental Health Sciences, Research Triangle Park, N. C. 27709, 1988.

39. Portier, C. J., Statistical properties of a two-stage model of carcinogenesis, Env. Health Persp. **76** (1987), 125–131.
40. Schwab, M., Varmus, H. E. and Bishop, J. M., Human N-myc gene contributes to neoplastic transformation of mammalian cells in culture, Nature **316** (1985), 160–162.
41. Scotto, J. and Nam, J. M., Skin melanoma and seasonal patterns, Am. J. Epidemiol. (1980), 309–314.
42. Spandido, D. A. and Wilkie, N. M., Malignant transformation of early passage rodent cells by a single mutated human oncogene H-ras-1 from T 24 bladder carcinoma line, Nature **310** (1984), 469–475.
43. Swerdlow, A. J., Incidence of malignant melanoma of the skin in England and Wales and its relationship to sunshine, Br. Med. J. **3** (1979), 1324–1327.
44. Tan, W. Y., Multivariate filtered Posson processes and applications to multivariate stochastic modeling of mutagenicity and carcinogenesis, Utilitas Mathematica **26** (1984), 63–78.
45. Tan, W. Y., A stochastic Gompertz birth-death process, Statist. and Prob. Lett. **4** (1986), 25–28.
46. Tan, W. Y. and Gastardo, M. T. C., On the assessment of effects of environmental agents on cancer tumor development by a two-stage model of carcinogenesis, Math. Biosciences **73** (1985), 143–155.
47. Tan, W. Y. and Brown, C. C., A nonhomogeneous two-stage model of carcinogenesis, Math. Modelling **9** (1987), 631–642.
48. Tan, W. Y. and Singh, K., A mixed model of carcinogenesis-with applications to retinoblastoma, Math. Biosciences **98** (1990), 201–211.
49. Tan, W. Y. and Starr, T. B. (1989), "Estimating initiation onset and tumor onset in animal carcinogenicity experiments with sacrifices by using a two-stage model of carcinogenesis," This paper was presented at the IMS-ASA- Biometric Society meeting, March 19-22, 1989, Lexington, KY.
50. Tan, W. Y. and Brown, C. C., A stochastic model of carcinogenesis-multiple pathways involving two-stage models, Paper presented at the Biometric Society (ENAR) meeting March 17–19, 1986, Atlanta, GA.
51. Tan, W. Y. and Chen, C., A stochastic model of carcinogenesis–Multiple pathways, Chapter 31 of Mathematical Population

Dynamics, eds. O. Arino, D. E. Axelrod & M. Kimmel, Marcel Dekker Inc., 1990.

52. Van Ryzin, J., Quantitative risk assessment, Jour. Occup. Medicine **22** (1980), 321–326.

53. Van Ryzin, J., Low dose risk assessment for regulating carcinogens, Center for the Social Sciences Newsletter, Columbia University, N. Y. **4, No. 1** (1984), 1–5.

54. Weinberg, R. A., The action of oncogenes in the cytoplasm and nucleus, Science **230** (1985), 770–776.

55. Weinberg, R. A., Finding the anti-oncogene, Scientific Amercian **259** (1988), 44–51.

56. Weinberg, R. A., The genetic origin of human cancer, Cancer **61** (1988), 1963–1968.

57. Weinberg, R. A., Oncogenes, antioncogenes and the molecular bases of multistep carcinogenesis, Cancer Research **49** (1989), 3713–3721.

58. Yancopoulos, G. D., et al., N-myc can cooperate with ras to transform normal cells in culture, Proc. Nat. Acad. Sci USA **82** (1985), 5455–5459.

5

Some Mixed Models of Carcinogenesis

Knudson (25) noted in retinoblastoma that eye cancer could be initiated either in germline cells or in somatic cells but the further late event always occurred in somatic cells. Founds (17) noted that many cancers developed by several independent pathways. In these situations, different individuals in the populations may involve different pathways for the process of carcinogenesis. Klein and Klein (24) have provided evidence which suggests that the number of stages in carcinogenesis depends heavily on the cell's environmental conditions. All these results point to the importance of mixed models of carcinogenesis. These biological results prompted Tan (46) and Tan and Singh (48) to develop mixed models of carcinogenesis. In this chapter I shall proceed to develop some mathematical theories for various mixed models of carcinogenesis and indicate some possible applications.

In Section (5.1), some biological evidence supporting the mixed models of carcinogenesis in the population are presented. Based on the biological evidence given in Section (5.1), in Section (5.2) we propose a general mixed model in the population involving four groups of subpopulations in each of which a different carcinogenesis model is entertained. By using results of probability generating functions (PGF) given in Section (5.2), we derive in Section (5.3) the age-specific incidence functions

179

of tumors and derive in Section (5.4) some procedures for computing the probability distributions of tumors. Having derived mathematical theories for the general mixed model given in Section (5.2), in Section (5.5) we then consider three important special cases of mixed models. To illustrate the usefulness of mixed models for some human cancers, we proceed in Section (5.6) to fit some retinoblastoma incidence data from NCI by a mixed model of carcinogenesis. Finally in Section (5.7), some applications of mixed models in risk assessment of environmental agents are indicated.

5.1 BIOLOGICAL EVIDENCE

Consider a large population of individuals and suppose that the population is divided into k nonoverlapping subpopulations. Then for certain cancers, it often happens that different subpopulations may involve different carcinogenesis models. Biologically this can happen conceivably from any of the three types of cancer genes: The oncogenes (1,3,5,6,8,34-35,52,54-56), the antioncogenes (22,27,53) and the accessory cancer genes (22,26).

Mixed Models of One-Stage Models and Two-Stage Models Related to Antioncogenes

Examples of this type of cancer in human beings include, among others, retinoblastoma which involves an antioncogene *Rb* in chromosome 13q14 (10,11,14,18,20), Wilm's tumor which involves an antioncogene *Wm* in chromosome 11p13 (16,20,28,40,56) and colon cancer which involves an antioncogene Fap in chromosome 5q21 (4,44). In these cases, the MVK two-stage model provides a biologically supported mathematical description of cancer development. In this two-stage model, while the second event is always somatic, the first event may be either in germline cells, in which case it is hereditary, or in somatic cells, in which case it is nonhereditary. As an example, consider the childhood cancer retinoblastoma. Then, if the first event occurs in a germline cell, all stem cells at the time of birth are intermediate cells with genotype $+/rb$ (rb denotes the mutated allele of Rb and + the wild allele of Rb). In these cases, one stage is enough to develop the cancer phenotype. (The genotype is either $-/rb$ or rb/rb, see 10,11,20.) On the other hand, if the first event occurs in a somatic cell and if the first event has not occurred before birth, all stem cells at the time of birth are normal cells so that two stages are required to yield cancer phenotype with genotype $-/rb$ or rb/rb. Thus, it is anticipated that the population consists of at least one-stage models and two-stage models of carcinogenesis. For retinoblastoma, both hereditary and non-hereditary cases have been reported (11,20). Note that with very small probability the first event may also occur in somatic cells before birth. In this case, at the time t_0 of birth, the stem cells consist of both normal stem cells and initiated cells so that cancer tumors may arise either from a one-stage model or from a MVK two-stage model. Taking this into account, the population model is then a mixture of one-stage models, MVK two-stages models and multiple pathway models involving one-stage and two-stage models.

Mixed Models of Two-Stage Models Related to Oncogenes

Oncogenes are directly responsible for many cancers (1,3,5,6,8,23,34, 35,49,52,54,55). It has been demonstrated in Chapter One that there are two complementing groups of oncogenes: One class is referred to as the "immortalization class" including nuclear oncogenes myc, myb and viral oncogenes SV-40 large T, polyma large T and adenovirus E1A while the other class is referred to as the "transforming class" and includes viral oncogene polyma middle T and cellular cytoplasmic oncogenes ras, src, erbB, neu, fms, fes/fps, yes, mil/raf, mos and abl (8,23,30,35,52,54,55). It has been demonstrated that cotransfection of one oncogene from the immortalization class with an oncogene from the transforming class would lead to tumorigenic conversion of normal stem cells (8,12,19,23,30,34,35,41,52,54,55). Since oncogenes in the same class share common properties (6,8,30,34,35,52,54,55), if the genetic background of different individuals in the population dictates that different individuals may assume different two-stage pathways and/or two-stage pathways with different mutation rates and/or different cell proliferation rates, it is then anticipated that the population model of carcinogenesis is a mixture of different two-stage models.

Mixed Models of Multiple Pathway Models Involving One-Stage and Two-Stage Models Related to Oncogenes

As discussed in the previous section, while two-stages are required for most cancers related to oncogenes, there are rare occasions in which one stage is sufficient for carcinogenesis. For example, the SV40 large T antigen oncogene can be considered as a dual oncogene (9) having both immortalization and transformation functions; thus SV40 large T antigen alone is capable of inducing conversion to a malignant tumor cell of normal stem cells (9,32). Spandidos and Wilkie (43) have also shown that tumorigenic conversion of rodent cells can be achieved by a single activated human oncogene (ras) if it is linked to a strong promoter; also, as reported by Keath et al. (21), Bignami et al. (2), Detto, Weinberg and Ariza (13) and Land et al. (31), if the normal neighboring cells are killed or removed by cytotoxic drugs, then activation of one oncogene in some cases also seems to be sufficient for inducing tumorigenic conversion of normal stem cells. Taking into account hereditary cases of cancers and independent multiple pathways, one would expect in the population a mixed model of multiple pathway models involving a one-stage model and a two-stage model of carcinogenesis. (See Section (4.5) of Chapter Four).

Mixed Models of Two-Stage Models Related to Accessory Cancer Genes

Mixtures of two-stage models arise also from accessory cancer genes which segregate in Mendelian fashion and are highly predisposable to certain cancers (22,26). Thus, unlike oncogenes, accessory cancer genes are genes which relate to cancers indirectly by increasing mutation rates of cancer genes (oncogenes or antioncogenes) and/or by facilitating cell proliferation of intermediate cells. Examples of accessory cancer genes in human beings include xeroderma pigmentosum (**xp**), ataxia telang-iectsia (**at**), Franconi's anemia (**fa**) and Bloom's syndrome. Consider for example the human inherited disease xeroderma pigmentosum for skin cancer. It has been shown (42) that for patients with the cancer prone gene xp, DNA repair for damages by ultraviolet light is defective, leading to an increased mutation rate for the initiation process in skin cancer. Let p be the frequency of the accessory cancer gene in question and assume random mating. Then, in the population, a proportion p^2 of individuals are homozygotes for this accessory cancer gene; a proportion $2p(1-p)$ of individuals are heterozygotes while a proportion $(1-p)^2$ of individuals do not contain the accessory gene. If the mutation rates and/or cell proliferation and differentiation rates are different for the three types of individuals, a mixture of three two-stage carcinogenesis models will result.

5.2 A GENERAL MIXED MODEL OF CARCINOGENESIS

Taking into account the biological evidence as given in the previous section, in this section I consider a situation in which there are $k = k_1 + k_2 + k_3 + k_4$ nonoverlapping subpopulations with proportion q_i for the *ith* population, $1 \leq i \leq k$; for the first group of k_1 populations, only one stage is required for carcinogenesis; for the next group of k_2 populations, two stages are required for carcinogenesis while for the third group of k_3 populations, cancer tumors may arise either by a one-stage model or by a two-stage model through a multiple pathway model; finally for the last group of k_4 populations, cancer may either arise from initiated cells by a one-stage model or arise from normal stem cells by a MVK two-stage model. A schematic presentation of this general mixed model of carcinogenesis is given in Figure 5.1.

In Figure 5.1, the one-stage models for the first group of k_1 populations arise basically from hereditary two-stage models in which the first event has occurred in a germline cell; in this case, therefore, all stem cells at birth (t_0) are initiated cells. The two-stage models in Figure 5.1 for the second group of k_2 populations correspond basically to the MVK two-stage models considered in Chapter Three; in this case, therefore, all stem cells at birth are normal stem cells. The one-stage and the two-stage multiple pathway models in Figure 5.1 for the third group of k_3 populations correspond basically to situations considered in Section (4.7) of Chapter Four; in this case, all stem cells at birth are normal stem cells. Finally for the last group of k_4 populations, the models arise basically from MVK two-stage models but with the first event having occurred before birth in somatic cells; in these cases, the stem cells at birth consist of both normal stem cells and initiated cells.

Let N, I_i and T denote a normal stem cell, an initiated cell in the *ith* population, $i = 1, \ldots, k$ and tumor, respectively. Let $N_i(t), I_i(t)$ and $T_i(t)$ be the numbers of N cells, I_i cells and cancer tumors, respectively, in the *ith* population, $i = 1, \ldots, k$ at time t. Let N_0 be the number of stem cells for the tissue in question at the time t_0 of birth. Then, for $i = 1, \ldots, k_1, N_i(t_0) = 0$ and $I_i(t_0) = N_0$; for $i = k_1 + 1, \ldots, k - k_4, N_i(t_0) = N_0$ and $I_i(t_0) = 0$; and for $i = k - k_4 + 1, \ldots, k, N_i(t_0) = N_{0(i)}$ and $I_i(t_0) = M_{0(i)}$, where $N_{0(i)} > 0$ and $M_{0(i)} > 0$ with $M_{0(i)} + N_{0(i)} = N_0$.

Since mutation rates are very small, one would expect that q_i for $1 \leq i \leq k_1$ and $k - k_4 < i \leq k$, would be very small and of the same

(1) One-Stage Models

$$I \xrightarrow{\beta_i(t)} T$$

(2) MVK Two-Stage Models

$$N \xrightarrow{\alpha_i(t)} I \xrightarrow{\beta_i(t)} T$$

(3) Multiple Pathway Models Involving One-Stage Models and Two-Stage Models

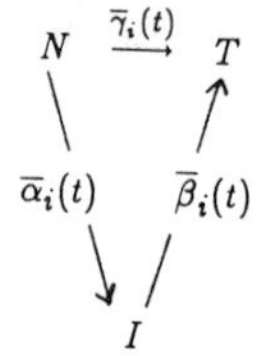

$$\gamma_i(t) = p_i(t)\overline{\gamma}_i(t), \quad \alpha_i(t) = [1 - p_i(t)]\overline{\alpha}_i(t),$$
$$\beta_i(t) = p_i(t)\overline{\gamma}_i(t) + [1 - p_i(t)]\overline{\beta}_i(t),$$

(see Section (4.5) of Chapter Four).

(4) A Mixture of a One-Stage and a Two-Stage Model

$$I \xrightarrow{\beta_i(t)} T$$

$$N \xrightarrow{\alpha_i(t)} I \xrightarrow{\beta_i(t)} T$$

(Stem cells consist of both normal stem cells and initiated cells.)

N=Normal Stem Cell, I=Initiated Cell, T=Cancer Tumor

Figure 5.1 A general mixed model involving one-stage, two-stage, a one-stage and two-stage multiple pathway and a mixture of one-stage and two-stage models.

order as the mutation rate of the first event. It follows that the second group and the third group of populations are not dominated by the other groups of populations.

5.2.1 Some Basic Assumptions

To derive mathematical theories for the mixed model in Figure 5.1, we make the following assumptions. Note that most of these assumptions are in fact the same assumptions made for the two-stage models in Chapter Three and the multiple pathway models in Chapter Four.

1. At the time t_0 of birth, the number N_0 of stem cells (normal or initiated stem cells) is very large ($\geq 10^6$). For the last group of k_4 populations, we assume that $N_{0(i)}$ is very large; for otherwise, the process would be dominated by the one-stage model so that the last group of populations could be included into the first group of populations. Since at the time of birth most organs are well developed, this assumption is expected to hold in most cases; see references (38) and (47).

2. In the ith population with $1 \leq i \leq k$, the intermediate cells (I cells) follow a nonhomogeneous Feller-Arley birth-death process with birth rate $b_i(t)$ and death rate $d_i(t)$ for cell proliferation and differentiation. Note that if $b_i(t) - d_i(t) = \epsilon_i exp(-\delta_i t)$ for some $\epsilon_i > 0$ and $\delta_i > 0$, then the process becomes a stochastic Gompertz process (45).

3. For $k_1 < i \leq k_1 + k_2$ and for $k - k_4 < i \leq k$, the probability that a normal stem cell at time t yields one normal stem cell and one I cell at time $t + \Delta t$ is $\alpha_i(t)\Delta t + o(\Delta t), \alpha_i(t) \geq 0$; similarly, for $1 \leq i \leq k_1 + k_2$ and for $k - k_4 < i \leq k$, the probability that an I cell at time t yields one I cell and one tumor cell at time $t + \Delta t$ is $\beta_i(t)\Delta t + o(\Delta t), \beta_i(t) \geq 0$.

4. For $k_1 + k_2 + 1 \leq i \leq k - k_4$, the process of carcinogenesis involves a multiple pathway model involving a one-stage model and a two-stage model. Thus, given the one-stage pathway from normal stem cells to tumor cells, let the probability that a normal stem cell at time t yields one normal stem cell and one tumor cell at time $t + \Delta t$ be $\bar{\gamma}_i(t)\Delta t + o(\Delta t)$, $\lim_{\Delta t \to 0} o(\Delta t)/\Delta t = 0$. Given the two-stage pathway $N \to I \to T$, let the probability that a normal stem cell at time t yields one normal stem and I cell at time $t + \Delta t$ be $\bar{\alpha}_i(t)\Delta t + o(\Delta t), \bar{\alpha}_i(t) \geq 0$, and let the probability that an I cell at time t yields one I cell and one tumor cell at time $t + \Delta t$ be $\bar{\beta}_i(t)\Delta t + o(\Delta t), \bar{\beta}_i(t) \geq 0$.

5. As in Moolgavkar and Knudson (38) and Tan and Brown (47), we assume that the time required for the development of tumors from tumor

cells is very short compared to the time required for the conversion of normal stem cells into cancer tumors. This implies that with probability one tumor cells grow into tumors and that one may ignore random variation for the time elapsed between the birth of tumor cells and the development of cancer tumors from tumor cells. Further, for human beings, the observed incidence rates are normally recorded for at least one year; in fitting incidence curves for human cancers, one may thus ignore the time elapsed between the birth of tumor cells and the development of cancer tumors from tumor cells; see references (38) and (47).

6.   The mutation processes are independent of the birth-death processes and all cells go through these processes independently of all other cells.

Given the above assumptions for the model in Figure 5.1, the following results then follow immediately:

1.   The processes $N_i(t), I_i(t)$ and $T_i(t)$ are basically Markov processes. This result is due to the fact that the future fate of any cell at time t depends only on the status of that cell at time t and is independent of its past history.

2.   Since the mutation rates are very small $(10^{-6}-10^{-8})$ and since cell proliferation rates of stem cells are usually nonnegative, under assumption (1) one would expect that $N_i(t)$ is very large for all $t \geq t_0$ and that $N_i(t)\gamma_i(t)$ for $k_1 + k_2 < i \leq k - k_4$ and $N_u(t)\alpha_u(t)$ for $k_1 < u \leq k$ are finite for all $t \geq t_0$. It follows that under assumption (1) one may ignore the randomness in $N_i(t)$ and assume some deterministic growth law for normal stem cells. Also, by the limiting theory of binomial distribution (law of small numbers), one would expect that for $k_1 < i \leq k_1 + k_2$ and for $k - k_4 < i \leq k$, the number of mutations from normal stem cells to I cells follows an approximately Poisson distribution with parameter $N_i(t)\alpha_i(t)\Delta t + o(\Delta t)$ during $[t, t + \Delta t]$ independently; similarly one would expect that for $k_1 + k_2 < i \leq k - k_4$, the number M_{1i} of mutations from normal stem cells to tumor cells and the number M_{2i} of mutations from normal stem cells to I cells follow independent Poisson distributions with parameters $N_i(t)\gamma_i(t)\Delta t + o(\Delta t)$ and $N_i(t)\alpha_i(t)\Delta t + o(\Delta t)$ during $[t, t + \Delta t]$, respectively. By the same argument, if N_0 is very large, then for $1 \leq i \leq k_1$, the number $I_i(t)$ of I cells is very large for all $t \geq t_0$ and that $I_i(t)\beta(t)$ is finite for all $t \geq t_0$. This is expected since the mutation rates $\beta_i(t)$ are usually very small $(\beta_i(t) \approx 10^{-6}-10^{-8}$ in most cases) and for $1 \leq i \leq k_1$, all stem cells are I cells and the cell proliferation rates of I cells are usually nonnegative. It follows that one may assume some de-

terministic growth law for the number of I cells of populations $1 \le i \le k_1$; further, by the limiting theory of binomial distribution (law of small numbers), the number M_i of mutations from I cells to tumor cells follows approximately a Poisson distribution with parameter $I_i(t)\beta_i(t) + o(\Delta t)$ during $[t, t + \Delta t]$, independently for $1 \le i \le k_1$.

3. From Tan and Brown (47), it is seen that the processes of $I_j(t)$ and $T_j(t)$ in the jth population follow continuous time multiple branches processes with transition probabilities during $[t, t + \Delta t]$ as given in Table (5.1), $j = 1, \ldots, k$. Further if, the normal stem cells follow a nonhomogeneous Feller-Arley birth-death process with birth rate $b_N(t)$ and death rate $d_N(t)$, then the processes $N_j(t), I_j(t)$ and $T_j(t)$ follow continuous time multiple branching processes with transition probabilities during $[t, t + \Delta t]$ as given in Tables (5.1) and (5.2), in the jth population, $j = 1, \ldots, k$.

4. For $k_1 + k_2 + 1 \le i \le k - k_4$, let $p_i(t)$ be the probability at time t for the one-stage pathway and put $\gamma_i(t) = p_i(t)\bar{\gamma}_i(t)$, $\alpha_i(t) = [1 - p_i(t)]\bar{\alpha}_i(t)$ and $\beta_i(t) = p_i(t)\bar{\gamma}_i(t) + [1 - p_i(t)]\bar{\beta}_i(t)$. Then, by assumption (d), the probability that a normal stem cell at time t yields one normal stem cell and one tumor cell at time $t + \Delta t$ is $\gamma_i(t)\Delta + o(\Delta t)$; the probability that a normal stem cell at time t yields one normal stem cell and one I cell at time $t + \Delta t$ is $\alpha_i(t)\Delta t + o(\Delta t)$, $\alpha_i(t) \ge 0$, while the probability that an I cell at time t yields one I cell and one tumor cell at time $t + \Delta t$ is $\beta_i(t)\Delta t + o(\Delta t)$, see also Section (4.7) of Chapter Four.

Table 5.1 Transitions and Transition Probabilities of I Cells and Tumors During $[t, t + \Delta t]$ in the jth Population.

Parent	Progeny	Probability
1 I Cell	2 I Cells	$b_j(t)\Delta t + o(\Delta t)$
1 I Cell	1 I Cell, 1 T Cell	$\beta_j(t)\Delta t + o(\Delta t)$
1 I Cell	Death	$d_j(t)\Delta t + o(\Delta t)$
1 I Cell	1 I Cell	$1 - [b_j(t) + d_j(t) + \beta_j(t)]\Delta t + o(\Delta t)$
1 T Cell	1 Tumor	$1 + o(\Delta t)$

Table 5.2 Transitions and Transition Probabilities of Normal Stem Cells During $[t, t + \Delta t]$ when $b_{i,N}(t) = ib_N(t)$ and $d_{i,N}(t) = id_N(t)$ in the jth Population.

Parent	Progeny	Probability
1 N Cell	2 N Cells	$b_N(t)\Delta t + o(\Delta t)$
1 N Cell	1 N Cell, 1 I Cell	$\alpha_j(t)\Delta t + o(\Delta t)$
1 N Cell	1 N Cell, 1 T Cell	$\gamma_j(t)\Delta t + o(\Delta t)$ if $k_1 + k_2 < j < k - k_4$; 0 if $k_1 < j \leq k_1 + k_2$ and $k - k_4 < j \leq k$.
1 N Cell	Death	$d_N(t)\Delta t + o(\Delta t)$
1 N Cell	1 N Cell	$1 - [b_N(t) + d_N(t) + \alpha_j(t) + \gamma_j(t)]\Delta t + o(\Delta t)$ if $k_1 + k_2 < j \leq k - k_4$; $1 - [b_N(t) + d_N(t) + \alpha_j(t)]$ $\Delta t + o(\Delta t)$ if $k_1 < j \leq k_1 + k_2$ or $k - k_4 < j \leq k$.

5.2.2 The Probability Generating Functions (PGF)

Let $P_{i,j}(t)$ be the conditional probability of $T(t) = j$ given N_0 stem cells (normal cells or initiated cells) at time t_0 in the ith population and $W_i(j,k;t)$ the conditional probability of $[I(t) = j, T(t) = k]$ given one I cell at time t_0 in the ith population.

Let $Q_i(z; t_o, t) = Q_i(t_0, t)$ be the PGF of $T(t)$ given N_0 stem cells at t_o in the ith population. Let $\phi_i(y, z; t_o, t) = \phi_i(t_0, t)$ be the PGF of $I(t)$ and $T(t)$ given one I cell at t_0 in the ith population and put $g_i(t_o, t) = g_i(z; t_o, t) = \phi_i(1, z; t_o, t)$.

Then,

$$Q_i(t_0, t) = \sum_{j=0}^{\infty} z^j P_{i,j}(t)$$

and

$$\phi_i(t_o, t) = \sum_{j=0}^{\infty} \sum_{k=0}^{\infty} y^j z^k W_i(j, k; t), i = 1, \ldots, k.$$

To obtain $Q_i(t_0, t)$, let $X_i(t)$ be the expected number of normal stem cells at time t given $N_{0(i)}$ normal stem cells at time t_0 in the ith population. ($N_{0(i)} = N_0$ for $k_1 < i \leq k - k_4$ and $N_{0(i)} = N_{0(i)}$ for $k - k_4 < i \leq k$). Then $X_i(t) \cong N_i(t)$ if $N_i(t)$ is very large for all $t \geq t_0$. Similarly, under assumption (1), $I_i(t)$ for $1 \leq i \leq k_1$ is very large for all $t \geq t_0$. In what follows, we thus assume that $I_i(t)$ for $1 \leq i \leq k_1$ and $N_i(t)$ for $k_1 < i \leq k$ are deterministic functions of t for $t \geq t_0$, unless otherwise stated.

THEOREM (5.2.1) For $1 \leq i \leq k_1$, assume that $I_i(t)\beta_i(t)$ is finite for all $t \geq t_0$ and that during $[t, t + \Delta t]$, the number of mutations from I cells to tumor cells follows a Poisson distribution with parameter $I_i(t)\beta_i(t) + o(\Delta t)$, independently. Then, under assumptions (1), (2), (5) and (6), we have for $1 \leq i \leq k_1$:

$$Q_i(t_0, t) = exp\{(z - 1)N_0 \int_{t_0}^{t} exp[\int_{t_0}^{x} \epsilon_i(u)du]\beta_i(x)dx\}$$

$$\text{for } 1 \leq i \leq k_1 \quad (5.2.1)$$

where $\epsilon_i(t) = b_i(t) - d_i(t)$

Proof. To prove Theorem (5.2.1), partition the time interval $[t_0, t]$ by $L_j = [t_{j-1}, t_j), j = 1, \ldots, n - 1$ and $L_n = [t_{n-1}, t_n]$ with $t_j = t_0 + j\Delta t, j = 1, \ldots, n$ and $t = t_n$. Let M_{ij} be the number of mutations from I cells to tumor cells during the time interval $L_j, j = 1, \ldots, n$ in the ith population, $1 \leq i \leq k_1$. Then M_{ij} follows a Poisson distribution with parameter $I_i(t_{j-1})\beta_i(t_{j-1})\Delta t + o(\Delta t), 1 \leq i \leq k_1, j = 1, \ldots, n$, independently. By assumption (5), the conditional PGF $Q_{ip}(t_0, t)$ of $T(t)$ in the ith population $(1 \leq i \leq k_1)$ is then given by:

$$Q_{ip}(t_0, t) = \Pi_{j=1}^{n}exp\{(z - 1)I_i(t_{j-1})\beta_i(t_{j-1})\Delta t + o(\Delta t)\}$$

Since $I_i(t) = N_0 exp\{\int_{t_0}^{t} \epsilon_i(x)dx\}$ by assumption (2), where $\epsilon_i(t) = b_i(t) - d_i(t)$, by letting $\Delta t \to 0$ one has, with $t = t_0 + n\Delta t$:

$$Q_i(t_0, t) = \lim_{\Delta t \to 0} Q_{ip}(t_0, t) = exp\{(z - 1) \lim_{\Delta t \to 0} \sum_{j=1}^{n}$$

$$I_i(t_{j-1})\beta_i(t_{j-1})\Delta t + t \lim_{\Delta t \to 0} o(\Delta t)/\Delta t\}$$

$$= exp\{(z - 1) \int_{t_0}^{t} I_i(x)\beta_i(x)dx\} =$$

$$exp\{(z - 1)N_0 \int_{t_0}^{t} exp[\int_{t_0}^{x} \epsilon_i(y)dy]\beta_i(x)dx\}.$$

To obtain $Q_i(t_0, t)$ for $k_1 < i \leq k$, observe that by (2) of Section (5.2.1), one may assume that for $k_1 < i \leq k$, the number of mutations from normal stem cells to I cells during $[t, t + \Delta t]$ along the two-stage pathway follows a Poisson distribution with parameter $N_i(t)\alpha_i(t)\Delta t + o(\Delta t)$, independently; similarly, for $k_1 + k_2 < i \leq k - k_4$, the number of mutations from normal stem cells to the tumor cells during $[t, t + \Delta t]$ along the one-stage pathway follows a Poisson distribution with parameter $N_i(t)\gamma_i(t)\Delta t + o(\Delta t)$. By Theorem (3.2.3) of Chapter Three, one has:

$$Q_i(t_0, t) = \exp\{\int_{t_0}^{t} N_i(x)\alpha_i(x)[q_i(x, t) - 1]dx\}$$

$$\text{for } k_1 < i \leq k_1 + k_2; \quad (5.2.2)$$

By Theorem (4.7.1) of Chapter Four, one has:

$$Q_i(t_0, t) = \exp\{(z - 1) \int_{t_0}^{t} N_i(x)\gamma_i(x)dx$$

$$+ \int_{t_0}^{t} N_i(x)\alpha_i(x)[g_i(x, t) - 1]dx\} \quad (5.2.3)$$

for $k_1 + k_2 < i \leq k - k_4 = k_1 + k_2 + k_3$.

Also, for $k - k_4 < i \leq k$, one has by assumption (6):

$$Q_i(t_0, t) = [g_i(t_0, t)]^{M_{0(i)}}\exp\{\int_{t_0}^{t} N_i(x)\alpha_i(x)[g_i(x, t) - 1]dx\}. \quad (5.2.4)$$

To apply formulas (5.2.2)–(5.2.4), one would need the functional forms of $N_i(t)$ and $g_i(t_0, t) = \phi_i(1, z; t_0, t)$.

The Expected Number $N_i(t)$ of Normal Stem Cells

As shown in Chapter Three, the functional form of $N_i(t)$ depends on the growth pattern of the tissue in question. In human cancers, there are basically three types of growth patterns: The logistic growth function, the Gompertz growth function, the Gamma-type growth functions or the exponential growth functions.

1. For breast cancer of women, the growth pattern of stem cells follows basically a logistic growth curve (see 37,38). Hence,

$$N_i(t) = N_{0(i)}\exp\{\int_{t_0}^{t} \epsilon_i(x)dx\}\{1 - (N_{0(i)}/M_i)$$

$$+ (N_{0(i)}/M_i)\exp[\int_{t_o}^{t} \epsilon_i(x)dx]\}^{-1} \text{ for } 0 < N_i(t) \leq M_i.$$

2. For the lung and the colon, the growth pattern of stem cells follows basically a Gompertz growth curve (37). Hence,

$$N_i(t) = N_{0(i)} exp\{(\epsilon_i/\delta_i)[exp(-t_0\delta_i) - exp(-t\delta_i)]\}$$

for some $\epsilon_i > 0$ and $\delta_i > 0$. (See 45).

3. For retinoblastoma, Wilm's tumor and Hodgkin's disease, the growth pattern of stem cells follows basically a Gamma type curve (37,38,48). Hence,

$$N_i(t) = a_i N_{0(i)} t^{c_i-1} exp(-\epsilon_i t)$$

for some $c_i \geq 1, \epsilon_i > 0$ and $a_i > 0$.

The PGF $\phi_i(t_0, t)$

Since $g_i(t_0, t) = \phi_i(1, z; t_0, t), g_i(t_0, t)$ is available if $\phi_i(t_0, t)$ is available. Now, as shown in Theorem (3.2.2) of Chapter Three, $\phi_i(t_0, t)$ satisfies the following first order partial differential equations:

$$\frac{\partial}{\partial t} \phi_i(t_0, t) = \{(y - 1)[yb_i(t) - d_i(t)] + y(z - 1)\beta_i(t)\} \frac{\partial}{\partial y} \phi_i(t_0, t),$$

$$\phi_i(t_0, t_0) = y. \tag{5.2.5}$$

Also, as shown in Section (3.3) of Chapter Three, for most of the important special cases, solution of equation (5.2.5) is readily available. Specifically, for the following three cases, solution of equation (5.2.5) has been obtained in Section (3.3) of Chapter Three.

1. If $b_i(t) = b_i, d_i(t) = d_i$ and $\beta_i(t) = \beta_i$ are independent of t, then $\phi_i(s, t) = \phi_i(t - s) = \phi_i(y, z; s, t)$ is given by:

$$\phi_i(s, t) = \phi_i(t - s) = \phi_i(y, z; s, t)$$
$$= \{u_{2i}(y - u_{1i}) + u_{1i}(u_{2i} - y)exp[(u_{2i} - u_{1i})b_i(t - s)]\}$$
$$\times \{(y - u_{1i}) + (u_{2i} - y)exp[(u_{2i} - u_{1i})b_i(t - s)]\}^{-1}, \tag{5.2.6}$$

where $u_{2i} > u_{1i}$ are the two real roots of the quadratic equation:

$$b_i x^2 - [b_i + d_i + \beta_i - z\beta_i]x + d_i = 0.$$

2. If $d_i(t) = 0$ for all $t \geq t_0$, then

$$\phi_i(s, t) = y[\xi_i(z; s, t) - y\eta_i(z; s, t)]^{-1},$$

where $\xi_i(z; s, t) = exp\{\int_s^t [b_i(x) + (1 - z)\beta_i(x)]dx\}$

and $\eta_i(z;s,t) = \int_s^t b_i(x)\xi_i(z;s,x)dx$.

3. Suppose that $[t_0,t]$ is partitioned into m nonoverlapping subinter-vals $L_u = [t_{u-1},t_u), u = 1,\ldots,m-1$ and $L_m = [t_{m-1},t_m]$ with $t = t_m$ and that in $L_j, b_i(t) = b_{ij} = b_{i,j}, d_i(t) = d_{ij} = d_{i,j}$ and $\beta_i(t) = \beta_{ij} = \beta_{i,j}$. To ob-tain $\phi_i(s,t)$, let $\phi_{i,j}(s,t)$ be the solution in formula (5.2.6) with (b_i, d_i, β_i) being replaced by $(b_{i,j}, d_{ij}, \beta_{ij})$ respectively. Then as shown in Section (3.3.3) of Chapter Three, the solution $\phi_i(s,t)$ of equation (5.2.5) with $s = t_0$ is:

If $t_{m-1} \le s < t \le t_m$, then $\phi_i(s,t) = \phi_{i,m}(s,t)$;

if $t_{j-1} \le s < t_j, j = 1,\ldots,m-1$, then

$\phi_i(s,t) = \phi_{ij}\{\phi_{ij+1}(t_j,t),z;s,t_j\}$. (Note $\phi_{ij}(s,t) = \phi_{ij}(y,z;s,t)$.)

For detail, the readers are referred to Section (3.3.3) of Chapter Three.

5.3 THE AGE-SPECIFIC INCIDENCE FUNCTION OF TUMORS

Let $\lambda_j(t)$ be the incidence rate of tumors at time t given N_0 stem cells at time t_0 in the jth population. Then the incidence function $\lambda(t)$ of tumors in the population given N_0 stem cells at t_0 is

$$\lambda(t) = \sum_{j=1}^k q_j \lambda_j(t).$$

To obtain $\lambda_j(t)$, put : $u_j(s,t) = [\partial\phi_j(s,t)/\partial y]_{(y=1,z=0)}$.

Then, by Theorem (2.1.1) of Chapter Two and using (5.2.1)–(5.2.4), we have:

$$\lambda_j(t) = -Q_j'(0;t_0,t)/Q_j(0;t_0,t) = N_0 \exp\left\{ \int_{t_0}^t \epsilon_j(x)dx \right\} \beta_j(t)$$

for $1 \le j \le k_1$;   (5.3.1)

$$\lambda_j(t) = -Q_j'(0;t_0,t)/Q_j(0;t_0,t) = \beta_j(t) \int_{t_0}^t N_j(x)\alpha_j(x)u_j(x,t)dx$$

for $k_1 < j \le k_1 + k_2$; and   (5.3.2)

$$\lambda_j(t) = -Q_j'(0;t_0,t)/Q_j(0;t_0,t)d = N_j(t)\gamma_j(t)$$

$$+ \beta_j(t) \int_{t_0}^{t} N_j(x)\alpha_j(x)u_j(x,t)dx \text{ for } k_1 + k_2 < j \leq k - k_4. \tag{5.3.3}$$

To obtain $\lambda_j(t)$ for $k - k_4 < j \leq k$, observe that by formula (5.2.4) and (5.2.5)

$$Q'_j(t_0,t) = M_{0(j)}[\phi_j(1,0;t_0,t)]^{M_{0(j)}-1}exp\{\int_{t_0}^{t} N_j(x)\alpha_j(x)$$

$$\times [\phi_j(1,0;x,t) - 1]dx\}[\partial\phi_j(1,0;t_0,t)/\partial t] + Q_j(0;t_0,t)\{N_j(t)\alpha_j(t)$$

$$\times [\phi_j(1,0;t,t) - 1] + \int_{t_0}^{t} N_j(x)\alpha_j(x)[\partial\phi_j(1,0;x,t)/\partial t]dx\}$$

$$= Q_j(0;t_0,t) M_{0(j)}[\phi_j(1,0;t_0,t)]^{-1}[-\beta_j(t)u_j(t_0,t)]$$

$$- Q_j(0;t_0,t)\beta_j(t) \int_{t_0}^{t} N_j(x)\alpha_j(x)u_j(x,t)dx.$$

Hence, for $k - k_4 < j \leq k$, one has:

$$\lambda_j(t) = -Q'_j(0;t_0,t)/Q_j(0;t_0,t)$$

$$= \beta_j(t)\{M_{0(j)}u_j(t_0,t)[\phi_j(1,0;t_0,t)]^{-1}$$

$$+ \int_{t_0}^{t} N_j(x)\alpha_j(x)u_j(x,t)dx\}, \tag{5.3.4}$$

If $\beta_j(t) \approx 0$ and if $\epsilon_j(t)$ is not big ($\epsilon_j(t) \leq 0.1.$), then as shown in Section (3.4.1) of Chapter Three, $\phi_j(1,0;t_0,t) \approx 1$ and a close approximation to $u_j(x,t)$ is given by $exp[\int_x^t \epsilon_j(y)dy]$. Hence if $\beta_j(t)$ is very small for all $t \geq t_0$ ($\beta_j(t) \approx 10^{-6}$–$10^{-8}$ in most cases), and if $\epsilon_j(t)$ is not big for all $t \geq t_0$ ($\epsilon_j(t) \leq 0.1$ in most cases), then

$$\lambda_j(t) \cong \beta_j(t) \int_{t_0}^{t} N_j(x)\alpha_j(x)exp[\int_x^t \epsilon_j(y)dy]dx \tag{5.3.5}$$

for $k_1 < j \leq k_1 + k_2$;

$$\lambda_j(t) \cong N_j(t)\gamma_j(t) + \beta_j(t) \int_{t_0}^{t} N_j(x)\alpha_j(x)exp[\int_x^t \epsilon_j(y)dy]dx \tag{5.3.6}$$

for $k_1 + k_2 < j \le k - k_4$ and

$$\lambda_j(t) \cong \beta_j(t)\{M_{0(j)}exp[\int_{t_o}^t \epsilon_j(x)dx]$$
$$+ \int_{t_0}^t N_j(x)\alpha_j(x)exp[\int_x^t \epsilon_j(y)dy]dx\} \tag{5.3.7}$$

for $k - k_4 < j \le k$.

Combining (5.3.1), (5.3.5), (5.3.6) and (5.3.7), if $\beta_i(t)$ is very small ($\beta_i(t) \approx 10^{-6}$–$10^{-8}$ in most cases) and if $\epsilon_i(t)$ is not big ($\epsilon_i(t) \le 0.1$ in most cases) for all $t \ge t_0$,

$$\lambda(t) = \sum_{i=1}^k q_i \lambda_i(t) \cong \sum_{i=1}^{k_1} q_i N_0 \beta_i(t)u_i(t_0,t)$$
$$+ \sum_{i=k_1+1}^{k_1+k_2} q_i\beta_i(t)\int_{t_0}^t N_i(x)\alpha_i(x)u_i(x,t)dx$$
$$+ \sum_{i=k-k_4}^{k} q_i\beta_i(t)\{M_{0(i)}u_i(t_0,t) + \int_{t_0}^t N_i(x)\alpha_i(x)u_i(x,t)dx\}$$
$$+ \sum_{i=k_1+k_2+1}^{k-k_4} q_i\{N_i(t)\gamma_i(t) + \beta_i(t)\int_{t_0}^t N_i(x)\alpha_i(x)u_i(x,t)dx\}$$
$$= \left\{\sum_{i=1}^{k_1} N_0 + \sum_{i=k-k_4}^{k} M_{0(i)}\right\} q_i\beta_i(t)u_i(t_0,t) \tag{5.3.8}$$
$$+ \sum_{i=k_1+1}^{k} q_i\beta_i(t)\int_{t_0}^t N_i(x)\alpha_i(x)u_i(x,t)dx + \sum_{i=k_1+k_2+1}^{k-k_4} q_i N_i(t)\gamma_i(t),$$

where $u_i(x,t) = exp\{\int_x^t \epsilon_i(y)dy\}$.

If $\alpha_i(t) = \alpha(t)$, $\epsilon_i(t) = \epsilon(t)$ and if the normal stem cells follow a nonhomogeneous Feller-Arley birth-death process for their proliferation with birth rate $b_N(t)$ and death rate $d_N(t)$, then $N_i(x) = N_{0(i)} exp\{\int_{t_0}^x \epsilon_N(y)dy\}$ and

(5.3.8) reduces to:

$$\lambda(t) \cong \Delta_1(t)u_I(t_0,t) + \Delta_2(t)u_N(t_0,t)$$
$$+ \Delta_3(t) \int_{t_0}^{t} \alpha(x)u_N(t_0,x)u_I(x,t)dx, \quad (5.3.9)$$

where

$$\Delta_1(t) = \left\{ \sum_{i=1}^{k_1} N_0 + \sum_{k-k_4}^{k} M_{0(i)} \right\} q_i\beta_i(t),$$

$$\Delta_2(t) = N_0 \sum_{i=k_1+k_2+1}^{k-k_4} q_i\gamma_i(t),$$

and

$$\Delta_3(t) = \sum_{i=k_1+1}^{k} q_i\beta_i(t)N_{0(i)},$$

and where

$$u_I(t_0,t) = exp\{ \int_{t_0}^{t} \epsilon(x)dx \} \quad \text{and} \quad u_N(t_0,t) = exp\{ \int_{t_0}^{t} \epsilon_N(x)dx \}.$$

Note that in (5.3.9), the first term corresponds to an incidence function of a one-stage model from intermediate cells to tumor cells; the second term corresponds to an incidence function of a one-stage model from normal stem cells to tumor cells while the last term corresponds to an incidence function of a MVK two-stage model considered in Chapter Three. Note that $\epsilon_i(t) = \epsilon(t)$ is equivalent to assuming that the cell proliferation rates of intermediate cells are the same for all populations. Similarly, $\alpha_i(t) = \alpha(t)$ is equivalent to assuming that the mutation rate from normal stem cells to intermediate cells are the same for all populations.

5.3.1 Some Special Cases

(a) If $\alpha_i(t) = \alpha_i$, $\beta_i(t) = \beta_i$, $\gamma_i(t) = \gamma_i$, $b_i(t) = b_i$ and $d_i(t) = d_i$ so that $\epsilon_i(t) = b_i - d_i = \epsilon_i$, then (5.3.8) reduces to:

$$\lambda(t) \cong \left\{ \sum_{i=1}^{k_1} N_0 + \sum_{i=k-k_4}^{k} M_{0(i)} \right\} q_i \beta_i exp[\epsilon_i(t-t_0)] \qquad (5.3.10)$$

$$+ \sum_{i=k_1+1}^{k} q_i \beta_i \alpha_i \int_{t_0}^{t} N_i(x) exp[\epsilon_i(t-x)]dx + \sum_{i=k_1+k_2+1}^{k-k_4} q_i \gamma_i N_i(t).$$

If further $\epsilon_i = \epsilon$ and $\alpha_i = \alpha$ and if the normal stem cells follow a homogeneous Feller-Arley birth-death process with birth rate b_N and death rate d_N for cell proliferation, then $N_i(x) = N_{0(i)} exp[\epsilon_N(x-t_0)]$ with $\epsilon_N = b_N - d_N$ and (5.3.10) reduces to:

$$\lambda(t) \cong \Delta_1 exp[\epsilon(t-t_0)] + \Delta_2 exp[\epsilon_N(t-t_0)]$$

$$+ \alpha\Delta_3 \int_{t_0}^{t} exp[\epsilon_N(x-t_0) + \epsilon(t-x)]dx$$

$$= \theta_1 exp[\epsilon(t-t_0)] + \theta_2 exp[\epsilon_N(t-t_0)], \qquad (5.3.11)$$

where $\Delta_i(t) = \Delta_i$, $\theta_1 = \Delta_1 + \alpha\Delta_3(\epsilon-\epsilon_N)^{-1}$ and $\theta_2 = \Delta_2 + \alpha\Delta_3(\epsilon_N-\epsilon)^{-1}$.
Note that $\lambda(t)$ in (5.3.11) is a linear combination of $exp[\epsilon(t-t_0)]$ and $exp[\epsilon_N(t-t_0]$, with coefficients θ_i independent of t.

(b) Suppose that $[t_0,t]$ is partitioned into n nonoverlapping subintervals $L_j = [t_{j-1},t_j)$, $j = 1,\ldots,n-1$ and $L_n = [t_{n-1},t_n]$ with $t_j = t_0 + j\Delta t$ and $t = t_n$ and that in L_j, $\alpha_i(t) = \alpha_{ij}$, $\beta_i(t) = \beta_{ij}$, $\gamma_i(t) = \gamma_{ij}$, $b_i(t) = b_{ij}$ and $d_i(t) = d_{ij}$. Let $\tau_j = t_j - t_{j-1}$, and put:

$$m_{ij}(x) = exp(\epsilon_{ij}x) \text{ and } v_{ij}(t_{j-1},t_j) = \int_{t_{j-1}}^{t_j} N_i(x) exp[\epsilon_{ij}(t_j-x)]dx.$$

Then, with $t = t_n$,

$$exp\{ \int_{t_0}^{t} \epsilon_i(x)dx \} = exp\left\{ \sum_{j=1}^{n} \int_{t_{j-1}}^{t_j} \epsilon_i(x)dx \right\} = \prod_{j=1}^{n} m_{ij}(\tau_j), \qquad (5.3.12)$$

and

$$\int_{t_0}^{t} N_i(x)\alpha_i(x) exp[\int_{x}^{t} \epsilon_i(y)dy]dx$$

$$= \sum_{j=1}^{n} \int_{t_{j-1}}^{t_j} N_i(x)\alpha_i(x) \, exp[\int_x^t \epsilon_i(y)dy]dx$$

$$= \sum_{j=1}^{n} \alpha_{ij}\nu_{ij}(t_{j-1},t_j) \prod_{u=j+1}^{n} m_{iu}(\tau_u). \tag{5.3.13}$$

On substituting (5.3.12) and (5.3.13) into $\lambda(t)$ of (5.3.8), (5.3.8) reduces to:

$$\lambda(t) \cong \left\{ \sum_{i=1}^{k_1} N_0 + \sum_{i=k-k_4}^{k} M_{0(i)} \right\} q_i \beta_{in} \prod_{j=1}^{n} m_{ij}(\tau_j)$$

$$+ \sum_{i=k_1+1}^{k} q_i \beta_{in} \sum_{j=1}^{n} \alpha_{ij}\nu_{ij}(t_{j-1},t_j) \prod_{u=j+1}^{n} m_{iu}(\tau_u)$$

$$+ \sum_{i=k_1+k_2+1}^{k-k_4} q_i \gamma_{in} N_i(t). \tag{5.3.14}$$

If $\alpha_i(t) = \alpha(t), \epsilon_i(t) = \epsilon(t)$ so that $\alpha_{ij} = \alpha_j$ and $\epsilon_{ij} = \epsilon_j$ are independent of i and if the normal stem cells follow a nonhomogeneous Feller-Arley birth-death process with rate $b_N(t)$ and death rate $d_N(t)$ for cell proliferation, then $m_{ij}(\tau_j) = exp(\epsilon_j\tau_j) = m_j(\tau_j)$ and

$$\nu_{ij}(t_{j-1},t_j) = \int_{t_{j-1}}^{t_j} N_i(x) exp[\epsilon_j(t_j - x)]dx$$

$$= N_{o(i)} \int_{t_{j-1}}^{t_j} exp[\int_{t_0}^x \epsilon_N(y)dy + \epsilon_j(t_j - x)]dx$$

$$= N_{0(i)}\nu_j(t_{j-1},t_j),$$

where $\nu_j(t_{j-1},t_j) = \int_{t_{j-1}}^{t_j} exp[\int_{t_0}^x \epsilon_N(y)dy + \epsilon_j(t_j - x)]dx$.
It follows that (5.3.14) reduces to:

$$\lambda(t) \cong \Delta_1(t_n) \prod_{j=1}^{n} m_j(\tau_j) + \Delta_2(t_n)exp[\int_{t_0}^t \epsilon_N(x)dx]$$

$$+ \Delta_3(t_n) \sum_{j=1}^{n} \alpha_j\nu_j(t_{j-1},t_j) \prod_{u=j+1}^{n} m_u(\tau_u). \tag{5.3.15}$$

Note that with $\Delta_3(t_n)$ replacing by $N_0\beta_n$, the third term in (5.3.15) is the same as that of formula (9) in Tan and Brown (47).

5.4   THE PROBABILITY DISTRIBUTION OF THE NUMBER OF CANCER TUMORS

Let $P_u(t)$ be the probability of having u tumors at time t given N_0 stem cells at time t_0 for an individual drawn randomly from the population. Let $P_{i,u}(t)$ be the probability of having u tumors at time t given N_0 stem cells at time t_0 for an individual from the ith population. Then

$$P_u(t) = \sum_{i=1}^{k} q_i P_{i,u}(t). \tag{5.4.1}$$

From (5.2.1), we have that for $1 \le i \le k_1$,

$$P_{i,u}(t) = exp[-\sigma_i(t_0,t)][\sigma_i(t_0,t)]^u/(u!), u = 0, 1, 2, \ldots, \tag{5.4.2}$$

where

$$\sigma_i(t_0,t) = N_0 \int_{t_0}^{t} exp[\int_{t_0}^{x} \epsilon_i(u)du]\beta_i(x)dx.$$

To obtain $P_{i,u}(t)$ for $k_1 < i \le k$, we expand $g_i(x,t)$ by Taylor series with respect to z around $z = 0$ to obtain:

$$g_i(x,t) = \sum_{j=0}^{\infty} z^j \omega_{i,j}(x,t), \tag{5.4.3}$$

where

$$\omega_{i,j}(x,t) = [\partial^j g_i(x,t)/\partial z^j]_{(z=0)}/(j!);$$

further, we write $exp\{\int_{t_0}^{t} N_i(x)\alpha_i(x)[q_i(x,t) - 1]dx$ as

$$exp\{\int_{t_0}^{t} N_i(x)\alpha_i(x)[g_i(x,t) - 1]dx\}$$

$$= exp\{-q_{i,0}(t_0,t) + \sum_{j=1}^{\infty} z^j q_{i,j}(t_0,t)\}$$

$$= \sum_{j=0}^{\infty} z^j \overline{P}_{ij}(t)/(j!), \tag{5.4.4}$$

where

$$q_{i,j}(t_0,t) = (-1)^{\delta_{oj}} \int_{t_0}^{t} N_i(x)\alpha_i(x)[\omega_{i,j}(x,t) - \delta_{oj}]dx$$

with δ_{oj} being the Kronecker's δ.

By Theorem (2.1.2) of Chapter Two,

$$\overline{P}_{i,0}(t) = exp[-q_{i,0}(t_0,t)]$$

and for $j \geq 1$

$$\overline{P}_{i,j}(t) = q_{i,j}(t_0,t)\overline{P}_{i,0}(t) + \sum_{u=1}^{j-1} q_{i,j-u}(t_0,t)\overline{P}_{i,u}(t)[(j-u)/j].$$

It follows that for $k_1 < i \leq k_1 + k_2$,

$$P_{i,j}(t) = \overline{P}_{i,j}(t), j = 0,1,2,\ldots \tag{5.4.5}$$

To obtain $P_{i,j}(t)$ for $k_1 + k_2 < i \leq k$, we let

$$V_{i,j}(t) = [\theta_i(t_0,t)]^j exp[-\theta_i(t_0,t)]/(j!),$$

where

$$\theta_i(t_0,t) = \int_{t_0}^{t} N_i(x)\gamma_i(x)dx;$$

and

$$[g_i(t_o,t)]^{M_{0(i)}} = \sum_{j=0}^{\infty} z^j U_{i,j}(t)/(j!), \tag{5.4.6}$$

where $U_{i,j}(t) = \{\partial^j [g_i(t_0,t)]^{M_{0(i)}}/\partial z^j\}_{(z=0)}$.

Then, from (5.2.4), for $k - k_4 < i \leq k$,

$$P_{i,j}(t) = \sum_{v=0}^{j} U_{i,v}(t)\overline{P}_{i,j-v}(t), j = 0,1,2\ldots; \tag{5.4.7}$$

and from (5.2.3), for $k_1 + k_2 < i \leq k - k_4$,

$$P_{i,j}(t) = \sum_{\ell=0}^{j} V_{i,\ell}(t)\overline{P}_{i,j-\ell}(t), j = 0,1,2\ldots \tag{5.4.8}$$

From (5.4.3), (5.4.4) and (5.4.6), note that to obtain $P_{i,j}(t)$ by using (5.4.5), (5.4.7) and (5.4.8), one would need $q_i(t_0,t)$. As shown in Section

(3.3) of Chapter Three and Section (5.2), for the three important special cases as considered in Section (5.2), $g_i(t_0,t)$ is available.

5.5 SOME SPECIAL CASES

In Section (5.2), we have presented a general mixed model consisting of four different groups of populations. In this section we give some specific models which arise from many practical situations.

Mixed Models of Carcinogenesis Involving an Antioncogene

As we have demonstrated before, many cancers are caused by inactivation of some antioncogenes. Well-known examples include retinoblastoma which involves the antioncogene Rb in chromosome 13q14 and Wilm's tumor which involves the antioncongene Wm in chromosome 11p13 (10, 11, 14, 16, 18, 20, 28 40, 56). In these cases, the population consists of three subpopulations, one subpopulation involving a one-stage model (hereditary cases), one subpopulation involving a two-stage model (nonhereditary case with the first event occurring after birth) and one subpopulation involving a multiple pathway model of a one-stage and a two-stage model (nonhereditary case but with the first event occurring in somatic cells before birth). In terms of the general model given in Section (5.2), $k_1 = k_2 = k_4 = 1$ and $k_3 = 0$.

Let p_i be the proportion of the ith population, $i = 1, 2, 3$ where $i = 3$ corresponds to the fourth population. If $\beta_i(t) \approx 0$ and if $\epsilon_i(t)$ is not big, then to the order $O(N_{0(i)}^{-1})$, the incidence function $\lambda(t)$ is approximately given by

$$\lambda(t) \cong \{N_0 p_1 \beta_1(t) u_1(t_0,t) + M_{0(3)} p_3 \beta_3(t) u_3(t_0,t)\}$$

$$+ \sum_{i=2}^{3} p_i \beta_i(t) \int_{t_0}^{t} N_i(x) \alpha_i(x) u_i(x,t) dx, \qquad (5.5.1)$$

where $u_i(x,t) = exp[\int_{x}^{t} \epsilon_i(y) dy], i = 1, 2, 3.$

If $\alpha_i(t) = \alpha_i, \beta_i(t) = \beta_i, b_i(t) = b_i$ and $d_i(t) = d_i$ so that $\epsilon_i(t) = b_i - d_i = \epsilon_i$, then (5.5.1) reduces to:

$$\lambda(t) \cong \{N_0 p_1 \beta_1 exp[\epsilon_1(t - t_0)] + M_{0(3)} p_3 \beta_3 exp[\epsilon_3(t - t_0)]\}$$

$$+ \sum_{i=2}^{3} p_i \beta_i \alpha_i \int_{t_0}^{t} N_i(x) \, exp[\epsilon_i(t-x)]dx \qquad (5.5.2)$$

Further, if the normal stem cells follow a homogeneous birth-death process with birth rate b_N and death rate d_N and if $\epsilon_i = \epsilon$ for $i = 1, 2, 3$, then $N_i(x) = N_{0(i)} \, exp[(x - t_0)\epsilon_N]$ with $\epsilon_N = b_N - d_N$ and (5.5.2) reduces to:

$$\lambda(t) \cong [N_0 p_1 \beta_1 + M_{0(3)} p_3 \beta_3] \, exp[\epsilon(t - t_0)]$$

$$+ [N_0 p_2 \alpha_2 \beta_2 + N_{0(3)} p_3 \alpha_3 \beta_3] \int_{t_0}^{t} exp[\epsilon_N(x - t_0) + \epsilon(t - x)]dx$$

$$= \theta_{11} \, exp[\epsilon(t - t_0)] + \theta_{12} \, \{exp[\epsilon(t - t_0)] - exp[\epsilon_N(t - t_0)]\} \quad (5.5.3)$$

where $\theta_{11} = N_0 p_1 \beta_1 + M_{0(3)} p_3 \beta_3$ and $\theta_{12} = [N_0 p_2 \alpha_2 \beta_2 + N_{0(3)} p_3 \alpha_3 \beta_3](\epsilon - \epsilon_N)^{-1}$.

In (5.5.3), observe that $\lambda(t)$ is a linear combination of $exp[\epsilon(t - t_0)]$ and $exp[\epsilon_N(t - t_0)]$. Since normal stem cells become immortalized by losing differentiation capability (7, 33, 36, 39), it is expected that $\epsilon \geq \epsilon_N$ so that $\theta_{12} \geq 0$. Since $\theta_{11} \geq 0$, the coefficient $\theta_{11} + \theta_{12}$ of $exp[\epsilon(t - t_0)]$ is ≥ 0 while the coefficient of $exp[\epsilon_N(t - t_0)]$ is $-\theta_{12} \leq 0$.

Note that the assumption $\epsilon_i(t) = \epsilon$ is equivalent to assuming that the cell proliferation rates of intermediate cells are the same and are independent of time for all populations.

Mixed Models of Carcinogenesis Involving Multiple Pathway Models

In cancers involving oncogenes and changing environments, it is expected that the population would consist of several subpopulations each of which involves a multiple pathway model of one-stage and two-stage carcinogenesis models. In terms of the general mixed model given in Section (5.2), $k_1 = k_2 = k_4 = 0$ and $k_3 = k \geq 1$.

Let p_i be the proportion of the ith subpopulation. If $\beta_i(t)$ is very small and if $\epsilon_i(t)$ is not big (≤ 0.1) for all $t \geq t_0$, then $\lambda(t)$ is approximately given by:

$$\lambda(t) \cong \sum_{i=1}^{k} p_i \{N_i(t)\gamma_i(t) + \beta_i(t) \int_{t_0}^{t} \alpha_i(x)N_i(x) \, exp[\int_{x}^{t} \epsilon_i(y)dy]dx\}.$$

$$(5.5.4)$$

If $\alpha_i(t) = \alpha_i, \beta_i(t) = \beta_i, \gamma_i(t) = \gamma_i, \epsilon_i(t) = \epsilon_i$ and if the normal stem cells follow a homogeneous Feller-Arley birth-death process with birth rate b_N and death rate d_N, then (5.5.4) reduces to:

$$\lambda(t) \cong \sum_{i=1}^{k} N_0 p_i \{\gamma_i exp[\epsilon_N(t - t_0)]$$

$$+\alpha_i \beta_i \int_{t_0}^{t} exp[\epsilon_N(x - t_0) + \epsilon_i(t - x)]dx\}, \text{ where } \epsilon_N = b_N - d_N. \quad (5.5.5)$$

If further $\epsilon_i = \epsilon$ for $i = 1, \ldots, k$, then

$$\lambda(t) \cong \theta_{21} exp[\epsilon_N(t - t_0)] + \theta_{22}\{exp[\epsilon(t - t_0)] - exp[\epsilon_N(t - t_0)]\}, \quad (5.5.6)$$

where $\theta_{21} = N_0 \sum_{i=1}^{k} p_i \gamma_i$, and $\theta_{22} = N_0(\epsilon - \epsilon_N)^{-1}(\sum_{i=1}^{k} p_i \alpha_i \beta_i)$.

From (5.5.6), we observe again that $\lambda(t)$ is a linear combination of $exp[\epsilon(t - t_0)]$ and $exp[\epsilon_N(t - t_0)]$. The condition $\epsilon \geq \epsilon_N$ implies that $\theta_{22} \geq 0$. Hence, when $\epsilon \geq \epsilon_N$, the coefficient θ_{22} of $exp[\epsilon(t - t_0)]$ is ≥ 0 while the coefficient $\theta_{21} - \theta_{22}$ of $exp[\epsilon_N(t - t_0)]$ may either be positive or negative depending on whether $\theta_{21} \geq \theta_{22}$.

Formula (5.5.3) and (5.5.6) suggest that for the homogeneous models with no selection, both the above two models give the same type of incidence functions. Thus, if one fits an observed incidence function to the theoretical ones, both the above two models should give identical results. Comparing (5.5.3), (5.5.6) with (5.3.11), the general model of Section (5.2) should also give the same result for the homogeneous models with no selection.

Mixed Models of Carcinogenesis Involving Oncogenes and/or Accessory Genes

In cancers involving oncogenes and/or accessory genes, under normal environmental conditions one would expect a mixed model of MVK two-stage models. For example, in the inherited human disease xeroderma pigmentosum, there are three genotypes involving the xeroderma pigmentosum gene (**xp**), **xp/xp**, +/**xp**, and +/ + . Since the mutation rates of the first event are different for different genotypes, it is anticipated that the population consists of three subpopulations in each of which a MVK two-stage model is involved. In terms of the general mixed model of Section (5.2), $k_1 = k_3 = k_4 = 0$ and $k_2 = k \geq 3$.

Let p_i be the proportion of the *ith* subpopulation. If $\beta_i(t)$ is very small and if $\epsilon_i(t)$ is not big (≤ 0.1) for all $t \geq t_o$, then $\lambda(t)$ is approximately given by:

$$\lambda(t) \cong \sum_{i=1}^{k} p_i \beta_i(t) \int_{t_0}^{t} N_i(x) \alpha_i(x) exp[\int_{x}^{t} \epsilon_i(y) dy] dx. \qquad (5.5.7)$$

If $\epsilon_i(y) = \epsilon(y), \alpha_i(t) = \alpha_i$ and if the cell proliferation rates and cell differentiation rates of normal stem cells are the same for all subpopulations, then $N_i(x) = N(x)$ and (5.5.7) reduces to:

$$\lambda(t) \cong \left(\sum_{i=1}^{k} p_i \alpha_i \beta_i(t)\right) \int_{t_0}^{t} N(x) exp[\int_{x}^{t} \epsilon(y) dy] dx. \qquad (5.5.8)$$

If further $\beta_i(t) = \beta_i$ so that $\sum_{i=1}^{k} p_i \alpha_i \beta_i$ is independent of t, then the shape of $\lambda(t)$ is determined mainly by $\int_{t_0}^{t} N(x) exp[\int_{x}^{t} \epsilon(y) dy] dx$. In these cases, the shape of the incidence function is exactly the same as that of a MVK two-stage model. Thus, in the inherited human disease xeroderma pigmentosum, if the **xp** gene does not affect cell proliferation and differentiation of initiated cells and if the mutation rates are time homogeneous, one may in fact adopt the Moolgavkar-Venzon-Knudson two-stage model to fit incidence functions.

5.6 FITTING OF RETINOBLASTOMA DATA BY A MIXED MODEL OF CARCINOGENESIS

It has been shown that retinoblastoma involves an antioncogene Rb in chromosome 13q14. To illustrate the fitting of incidence data by a mixed model, in this section we proceed to fit the observed incidence data of the SEER (Surveillance Epidemiology End Results) project of the NCI (National Cancer Institute) for the years 1973 to 1982 by a model given in Section (5.5). Given in columns 2 and 6 of Table (5.3) are the observed incidences per million per year which were obtained from the original SEER data by multiplying the original observed incidences per year by $M/10^6$, where M is the one year population size. These data were kindly supplied to the author by Dr. John Young of the NCI.

To fit these data, we assume that $\epsilon_i(t) = \epsilon, \epsilon_N(t) = \epsilon_N, \alpha_i(t) = \alpha_i, \beta_i(t) = \beta_i$ so that the incidence function is approximately given by (5.5.3). These assumptions do not appear to be very restrictive; see Moolgavkar and Knudson (38) and Moolgavkar (37). To fit (5.5.3) to the retinoblastoma data, observe that the scaling for the incidence is

Table 5.3 Observed and Predicteed Incidence: (1973–1982)

Sex	Male				Female			
	Observed Cases		Predicted		Observed Cases		Predicted	
Age t (years)	$\lambda(t)$	$\lambda_I(t)$	$\lambda_N(t)$	$\hat{\lambda}(t)$	$\lambda(t)$	$\lambda_I(t)$	$\lambda_N(t)$	$\hat{\lambda}(t)$
0	20.7677	20.9351	0.0000	20.9351	22.6548	22.6334	0.0000	22.6334
1	10.3834	11.2776	0.6107	11.8883	15.9423	11.7579	4.1840	15.9419
2	8.7859	6.0753	0.6180	6.6933	8.3907	6.1082	2.3706	8.4788
3	4.7923	3.2727	0.4697	3.7424	4.1953	3.1732	1.2408	4.4140
4	0.7987	1.7630	0.3178	2.0808	3.3563	1.6484	0.6450	2.2934
5	0.2856	0.9497	0.2017	1.1514	0.2242	0.8564	0.3351	1.1915
Parameter	ω_1	ω_2	$\bar{\epsilon}_N$	$\bar{\epsilon}_I$	ω_1	ω_2	$\bar{\epsilon}_N$	$\bar{\epsilon}_I$
Estimate	9.3144	20.9351	−.7484	−.6186	8.8568	22.6334	−3.0557	−0.6549
Std. Error	0.9233	1.5270	0.9974	0.9974	0.3642	0.65144	0.0093	0.1152

The goodness of fit for $\hat{\lambda}(t)$ is 2.5826 for the male population (p-value is 0.763) and is 1.2897 for the female population (p-value is 0.934).

"per million" and "per year" and that the average doubling time for human cells is approximately 28 days. Hence, in fitting the incidence data we multiply (5.5.3) by 13×10^6 and change ϵ and ϵ_N into $\bar{\epsilon} = 13\epsilon$ and $\bar{\epsilon}_N = 13\epsilon_N$ to obtain:

$$\lambda(t) \cong w_1 \exp[\bar{\epsilon}(t - t_0)] + w_2\{\exp[\bar{\epsilon}_N(t - t_0)] - \exp[\bar{\epsilon}(t - t_0)]\}, \quad (5.6.1)$$

where $w_1 = 13 \times 10^6 \theta_{11}$ and $w_2 = 13 \times 10^6 \theta_{12}$.

We use a nonlinear least square method to fit (5.6.1) to the retinoblastoma data by making use of the BMDP package.

Assuming equal weights, the estimates $(\hat{w}_1, \hat{w}_2.13\hat{\epsilon}, 13\hat{\epsilon}_N)$ of $(w_1, w_2, \bar{\epsilon}, \bar{\epsilon}_N)$ together with the estimates of standard errors of these estimates are given in Table (5.3) (see Remark 2). For assessing the fit, we also give in Table (5.3) the predicted incidences $\hat{\lambda}_I(t), \hat{\lambda}_N(t)$ and $\hat{\lambda}(t)$ per million per year and the goodness of fit statistics. Note from Table (5.3) that both values of $\hat{\epsilon}_N$ and $\hat{\epsilon}$ are negative. Thus, one would expect that the number of normal stem cells would eventually reduce to zero so that after several years almost all cells of the retina are differentiated structure cells. This may help explain why retinoblastoma rarely occurs after the age of five (see Table (5.3) for observed incidence).

From the results in Table (5.3), it is observed that the predicted incidences fit quite well with the observed incidences, especially for the female population; furthermore, the contribution to $\hat{\lambda}(t)$ appears to be mainly from the one-stage model although the contribution of the two-stage model also appears to be quite significant.

To evaluate the relative contribution of $\hat{\lambda}_N(t)$ and $\hat{\lambda}_I(t)$ to $\hat{\lambda}(t)$, we compute the ratio $\hat{\lambda}_N(t)/\hat{\lambda}(t)$ for various t values as given in Table (5.4).

From Table (5.4) one observes immediately that for the female population, the ratio stabilizes to .39 at $t = 2$, suggesting that approximately 60% of the incidence derives from a one-stage model.

To help explain results from Table (5.4), note that $\lambda_N(t)/\lambda_I(t) = (\theta_{12}/\theta_{11})\{1 - \exp[-(\epsilon - \epsilon_N)t]\}$ is an increasing function of t if $\epsilon > \epsilon_N$

Table 5.4

$\hat{\lambda}_N(t)/\hat{\lambda}_I(T)$ $t =$	1	2	3	4	5
Male	.05	.10	.14	.18	.21
Female	.36	.39	.39	.39	.39

and $\lambda_N(t)/\lambda_I(t) \to \theta_{12}/\theta_{11}$ as $t \to \infty$ if $\epsilon > \epsilon_N$; further the rate of convergence of the ratio depends on the difference $\epsilon - \epsilon_N$. Note that $\hat{\epsilon} > \hat{\epsilon}_N$ and $\epsilon > \epsilon_N$ is consistent with the hypothesis that normal stem cells become immortalized by losing their capability of differentiation.

REMARK There is no reason to believe that the usual variance estimates based on the usual least square method will hold for fitting incidence functions. To obtain the variance of the estimates, we thus use a bootstrap method (i.e. a stratified resampling scheme) due to Efron (15) to derive estimates of standard errors of estimates. In this method, for each fixed time, data were generated by computer for all individuals that die at the time by substituting the parameters by the estimates in the model. Then by sampling with replacement from within these strata, a new data set with the same design as the original data set is obtained. By repeating this process 200 times, the variances of the estimates of the model parameters were estimated.

5.7 SOME POSSIBLE APPLICATIONS OF MIXED MODELS OF CARCINOGENESIS

By comparing results from the one-hit model, multi-hit model, Weibull model and classical Armitage–Doll multistage model, Van Ryzin (50, 51) has shown that for risk assessment of carcinogens, different models give very different results. It follows that for obtaining correct risk assessment, the adopted model should truly reflect the biological mechanisms. In this sense then, the mixed models proposed in this paper should be useful for risk assessment of environmental agents. Clearly, more research is needed to develop statistical inference procedures for mixed models of carcinogenesis. Krailo, Thomas and Pike (29) have in fact assumed a mixed-type model involving the Moolgavkar-Venzon-Knudson two-stage model to analyze case-control breast cancer data of Los Angeles involving 8 time-dependent covariates. This result suggests that developing statistical inference procedures for mixed models might not be very difficult although lots of effort and hard work are required.

As in the case of multiple pathway models, one may also use the mixed models to evaluate the robustness and the efficiencies of risk assessment procedures for environmental agents. As described in Chapter Four, given the parameter values one may generate data by using mixed models. By this approach, Monte Carlo studies can be generated to evaluate the robustness and the efficiencies of risk assessment procedures for environmental agents.

5.8 CONCLUSIONS AND SUMMARY

Much biological evidence exists indicating that for many cancers the process of carcinogenesis in the population is a mixture of different carcinogenesis models. Based on the biological evidence and on careful examination of the functions of the three types of cancer genes-antioncogenes, oncogenes and accessory cancer genes, in this chapter we propose a mixed model of carcinogenesis in the population involving four groups of subpopulations in each of which a different carcinogenesis model is entertained.

To start this chapter, we present in Section (5.1) some biological evidence supporting the mixed models of some different carcinogenesis models. As illustrated in Section (5.1), for cancers involving only antioncogenes, one may expect a mixed carcinogenesis model of one-stage models, two-stage models and mixtures of a one-stage model and a two-stage model; conceivably these types of mixed models may occur because for the MVK two-stage models related to antioncogenes, the first event occurs either in germline cells or in somatic cells whereas the second event always occurs in somatic cells. For cancers involving oncogenes and/or accessory cancer genes, one may expect mixed models of some MVK two-stage models because of the recessive nature of the accessory cancer genes and because oncogenes in the same class share some common functions and properties; for cancers involving oncogenes only, because of the existence of dual oncogenes and the changing environmental conditions, one may also expect mixed models of multiple pathway models involving a one-stage model and a two-stage model.

For developing mathematical theories for the proposed mixed model, in Section (5.2) we derive the probability generating functions (PGF) of the number of tumors for each population under some general conditions. As shown in Section (5.2), for the one-stage models the carcinogenesis process follows a nonhomogeneous Poisson process whereas for the two-stage models the carcinogenesis process usually follows a nonhomogeneous filtered Poisson process. By using the PGF's of tumors given in Section (5.2) we derive in Section (5.3) the age-specific incidence function of tumors and give an approximation to the incidence function by assuming that the mutation rates are very small $(10^{-6}$–$10^{-8})$

and the cell proliferation rates are not big (≤ 0.1). Again by using the PGF's of tumors we derive in Section (5.4) an iterative procedure for computing the probabilities of tumors.

Having derived mathematical theories for the general mixed model given in Section (5.1), in Section (5.5) we consider three important special cases of mixed models in each of which the incidence function of tumors is derived. These special cases include: (i) mixed models of one-stage and two-stage models and mixtures of a one-stage and a two-stage model, which prevail in cancers related to anti-oncogenes, (ii) mixed models of two-stage models, which prevail in cancers related to accessory cancer genes because of the recessive nature of these genes and the Mendelian segregation of these genes, and to oncogenes because oncogenes in the same class share some common functions, and (iii) mixed models of multiple pathway models involving a one-stage model and a two-stage model, which prevail in cancers related to oncogenes because of the existence of dual oncogenes and the changing environment. It is shown in Section (5.5) that if all parameters are time homogeneous, if the cell proliferation rates of I cells are the same for all populations and if the normal stem cells follow a homogeneous birth-death process for all populations, then the incidence function of the general model given in Section (5.2) becomes a linear combination of $\exp\{\epsilon_N(t-t_0)\}$ and $\exp\{\epsilon(t-t_0)\}$, where ϵ_N and ϵ denote the proliferation rates of normal stem cells and I cells, respectively. To illustrate the applications of the mixed models, in Section (5.6) we illustrate how to fit some retinoblastoma incidence data of NCI by using a mixed model of one-stage models, two-stage models and mixtures of a one-stage model and a two-stage model. The excellent fit of the retinoblastoma data suggests the usefulness of the mixed model. Finally in Section (5.7), some possible applications of the mixed models in assessing risks of environmental agents are indicated.

REFERENCES

1. Barbacid, M., Ras genes, Annu. Rev. Biochem **56** (1987), 779–827.
2. Bignami, M. et al., Differential influence of adjacent normal cells on the proliferation of mammalian cells transformed by the viral oncogenes myc, ras, and src, Oncogene **2** (1988), 509–514.
3. Bishop, J.M., The molecular genetic of cancer, Science **235** (1987), 305-311.

4. Bodmer, W.F. et al., Localization of the gene for familial adenomatous polyposis on chromosome 5, Nature **328** (1984), 614-616.

5. Bos, J. L., Ras oncogenes in human cancer: A review, Cancer Res **49** (1989), 4682– 4689.

6. Buckley, I., Oncogenes and the nature of malignancy, Adv. in Cancer Res **50** (1988), 71–93.

7. Buick, R.N. and Pollak, M.N., Perspective on clonogenic tumor cells, stem cells and onocogenes, Cancer Res. **44** (1984), 4909-4918.

8. Burck, K. B., Liu, E. T. and Larrick, J. W., Oncogenes: An Introduction to the Concept of Cancer Genes (1988), Springer-Verlag, New York.

9. Butel, J.S., SV40 large T antigen: Dual oncogene, Cancer Survey **5** (1986), 343-366.

10. Cavenee, W.K. et al., Expression of recessive alleles by chromosomal mechanisms in retinoblastoma, Nature **305** (1983), 719-784.

11. Cavenee, W.K. et al., Genetic origin of mutations predisposing to retinoblastoma, Science **228** (1985), 501-503.

12. Compere, S. J. et al., The ras and myc oncogenes cooperate in tumor induction in many tissues when introduced into midgestation mouse embryos by retroviral vectors, Proc. Natl. Acad. Sci USA **86** (1989), 2224–2228.

13. Detto, G. P. Weinberg, R. A. and Ariza, A, Malignant transformation of mouse primary keratinocytes by HaSV and its modulation by surrounding normal cells, Proc. Natl. Acad. Sci. **86** (1988), 6389–6393.

14. Dryja, T.P. et al., Homozygosity of chromosome 13 in retinoblastoma, The New England J. Medicine **310** (1984), 550-553.

15. Efron, B., The Jacknife, the Bootstrap and other Resampling Plans, (1982), Philadelphia, SIAM.

16. Fearson, E.R. Volgestein, B. and Feinberg, A.P., Somatic deletion and duplication of genes on chromosome 11 in Wilm's tumors, Nature **309** (1984), 174-176.

17. Founds, L., Neoplastic Development **2** (1975), Academic Press, N.Y..

18. Fung, J.K.T. et al., Structural evidence for the authenticity of the human retinoblastoma gene, Science **236** (1987), 1657-1660.

19. Glaichenhaus, N. et al., Cooperation between multiple oncogenes in rodent embryo fibroblasts: An experimental model of tumor progression, Advances in Cancer Research **45** (1985), 291–305.

20. Hansen, M. F. and Cavene, W. K., Genetics of cancer predisposition, Cancer Res. **47** (1987), 5518–5527.

21. Keath, E.J., Caimi, P.G. and Cole, M.D., Fibroblast lines expressing activated c-myc oncogenes are tumorigenic in nude mice and syngeneic animals, Cells **39** (1984), 339-348.
22. Klein, G., The approaching era of the tumor suppressor genes, Science **238** (1987), 1539–1545.
23. Klein, G. and Klein, E., Oncogene activation and tumor progression, Carcinogenesis **5** (1984), 429-436.
24. Klein, G. and Klein E., Conditional tumorigenicity of activated oncogene, Cancer Res. **46** (1986), 3211-3224.
25. Knudson, A.G., Mutation and Cancer: Statistical study of retinoblastoma, Proc. Natl. Acad. Sciences USA **68** (1971), 820-823.
26. Knudson, A.G., Cancer genes in man, An International Colloquium. M.D. Anderson Tumor Hospital, Houston, TX Nov., 1981.
27. Knudson, A.G., Hereditary cancer, oncogenes and antioncogenes, Cancer Res. **45** (1985), 1437-1443.
28. Koufos, A. et al., Loss of alleles at loci on human chromosome 11 during genesis of Wilm's tumor, Nature **309** (1984), 170-172.
29. Krailo, M., Thomas, D. and Pike, M., Fitting Models of carcinogenesis to a case-control study of breast cancer, Symposium on "Time-Related Factors in Cancer Epidemiology", April 15-17, 1985. NCI/NIH, Bethesda, MD. Jour. Chronic Disease 40 Supplement 2 (1987).
30. Land, H., Parada, L.F. and Weinberg, R.A., Tumorigenic conversion of a primary embryo fibroblasts requires at least two cooperating oncogenes, Nature **304** (1983), 596-601.
31. Land, H., et al., Behavior of myc and ras oncogenes in transformation of rat embryo fibroblasts, Mol. Cell Biol. **6** (1986,), 1917- 1925..
32. Lanford, R.E., Wong, O., and Butel, J.S., Differential ability of a T-antigen transport-defective mutant of similar virus 40 to transform primary and established rodent cells, Molecular Cellular Biology **5** (1985), 1043-1050.
33. Mackillop, W.J., Ciampi, A., and Buick, R.N., A stem cell model of human tumor growth: Implications for tumor cell clonogenic assays, J. Natl. Cancer Inst. **70** (1983), 9-16.
34. Marshall, C. J., Oncogenes, J. Cell Sci. Suppl. **4** (186), 417–430.
35. Marshall, C.J. and Ridby, P.W.J.C., Viral and cellular genes involved in oncogenesis, Cancer Survey **3** (1984), 183-214.
36. Matsumura, T., Hayashi, M. and Konishi, R., Immortalization in culture of rat cells: A genealogic study, J. Natl. Cancer Inst. 74 (1985), 1223-1232.

37. Moolgavkar, S.H., Carcinogenesis modeling: From Molecular biology to epidemiology, Ann. Rev. Public Health **7** (1986), 151-169.
38. Moolgavkar, S.H. and Knudson, A.G., Mutation and cancer: A model for human carcinogenesis, J. Natl. Cancer Inst. **66** (1981), 1037-1052.
39. Oberley, L. W. and Oberley, T.D., The role of superoxide dismutase and gene amplification in carcinogenesis, J. Theor. Biology **106** (1984), 403- 422.
40. Orkin, S.H. et al., Development of homozygosity for chromosome 11p markers in Wilm's tumors, Nature **309** (1984), 172-174.
41. Sinn, E. et al., Coexpression of MMTV/v-Ha-ras and MMTV/c-myc genes in transgenic mice: Synergistic action of oncogenes in vivo, Cell **49** (1987), 465- 475.
42. Sirover, M.A. et al., Cellular and molecular regulation of DNA repair in normal human cell and in hypermutable cells from cancer prone individuals, In: "International Conference on Mechanisms of DNA Damage and Repair. June 2-7, 1985," Eds. M.G. Simic, L. Grossman & A.C. Upton, Plenum Press, N.Y., 1986.
43. Spandidos, D.A. and Wilkie, N.M., Malignant transformation of early passage rodent cells by a single mutated human oncongene H-ras-1 from T 24 bladder carcinoma line, Nature **310** (1984), 469-475.
44. Solomon, E. et al., Chromosome 5 allele loss in human colorectal carcinomas, Nature **328** (1987), 616-619.
45. Tan, W.Y., A stochastic Gompertz birth-death process, Statist. and Prob. Lett. **4** (1986), 25-28.
46. Tan, W.Y., Some mixed models of carcinogenesis, Math. Modelling 10(1988), 765-773..
47. Tan, W.Y. and Brown, C.C., A nonhomogeneous two-stage model of carcinogenesis, Math Modelling **9** (1987), 631-642.
48. Tan, W.Y. and Singh, K., A mixed model of carcinogenesis-with applications to retinoblastoma, Math. Biosciences **98** (1990), 201–211.
49. Topal, M. D., DNA repair, oncogenes and carcinogenesis, Carcinogenesis **9** (1988), 691–696.
50. Van Ryzin, J., Quantitative risk assessment, Jour. Occup. Medicine **22** (1980), 321- 326.
51. Van Ryzin, J., Low dose risk assessment for regulating carcinogens, Center for the Social Sciences Newsletter, Columbia University, NY **4 No. 1** (1984), 1- 5.
52. Weinberg R.A., The action of oncogenes in the cytoplasm and nucleus, Science **230** (1985), 770-776.

53. Weinberg, R. A., Finding the antioncogene, Scientific Amer. **259** (1988), 44–51.
54. Weinberg, R. A., The genetic origin of human cancer, Cancer **61** (1988), 1963– 1968.
55. Weinberg, R. A., Oncogenes, antioncogenes and the molecular bases of multistep carcinogenesis, Cancer Research **49** (1989), 3713–3721.
56. Weissman, B.E. et al., Introduction of a normal human chromosome 11 into a Wilm's tumor cell line controls its tumorigenic expression, Science **236** (1987), 175-180.

6

Multievent Models of Carcinogenesis

It has been demonstrated in Chapter One that carcinogenesis is a multistep random process with intermediate cells subjected to stochastic cell proliferation and differentiation. This consideration has let Chu (7), Tan, Chu and Brown (21) to extend the MVK two-stage model into a multistage model with cell proliferation and cell differentiation for normal stem cells and intermediate cells. Chu (7) has called his model the *"multievent"* model to distinguish it from the classical Armitage–Doll multistage model (1, 2). Chu et al. (8) have illustrated how to use their model to discriminate between different cancer promotion models. In this chapter I proceed to provide some mathematical theories of the multievent models and suggest some possible applications.

In Section (6.1), some biological evidence for multievent models are presented. Then in Section (6.2), multievent models are proposed and the probability generating functions (PGF) of the number of tumors for these models are derived. By using the PGFs of the number of tumors obtained in Section (6.2), we derive in Section (6.3) the incidence functions of the number of tumors and derive in Section (6.4) the expected values of the number of tumors and the variances and covariances of the numbers of tumors.

Having derived mathematical theories for the general multievent models, in Sections (6.5) and (6.6) we then consider two important special cases of multievent models. Finally in Section (6.7), some applications of multievent models are indicated.

6.1 BIOLOGICAL EVIDENCE

It has been demonstrated in Chapters One and Three that for cancers related to oncogenes, at least two events are required to convert normal stem cells into malignant tumors. These two events are the *"immortalization"* and *"transformation"* of stem cells. Finer analysis at the molecular level indicates that in some occasions, immortalization and transformation of stem cells are not enough for the malignant conversion of normal stem cells. Given below are some specific examples:

1. Land, Parada and Weinberg (11) reported that while cotransfection of activated *EJras* oncogene and *myc* oncogene induced tumors in nude mice, however, these induced tumors only grew to a maximum size of 2 *cm*, indicating that at least one more step was necessary to form malignant tumors.

2. If immortalization and transformation are the only two events required to convert normal stem cells into malignant tumors, then immortalized cell lines should be transformed by the oncogene *ras* or *src* alone from the transformation class. However, Ruley et al. (15), Tsunokawa et al. (25) and Thomassen et al. (24) have observed the existence of immortalized cell lines which cannot be transformed by the oncogene *ras* or *src* alone.

3. Barrett et al. (3) and Oshimura, Gilmer and Barrett (14) reported that hybrids between normal Syrian hamster embryo (SHE) cells and tumorigeneous SHE cells which were induced by $v-Ha-ras$ plus $v-myc$ oncogenes had lost their tumorigenicity; yet these hybrids still expressed the *ras* and *myc* RNA and high levels of the mutated form of the $p21^{ras}$ protein, indicating that the loss of tumorigenicity was not due to the loss or lack of expression of the oncogenes. When these hybrid cells were passaged, anchorage-independent and tumorigenic variant cells arose at a low frequency in the populations. Karyotype analysis indicated that hybrids which were suppressed for tumorigenicity and anchorage-independence contained normal karyotypes as normal cells; yet hybrids

which re-expressed tumorigenicity had a nonrandom loss of chromosome 15.

4. By introducing a retroviral vector containing both v-ras and v-myc into midgestation mouse embryos, Compere et al. (6) found that the *ras* and *myc* oncogenes cooperated to induce tumors in a wide variety of organs including brain, skin and kidney but the rates of tumor induction were quite different in different organs; also, while the majority of tumors in brain neoplasm, skin squamous cell carcinoma, kidney neoplasm and spindle cell neoplasms in heart, skin and subcutaneous tissue, are malignant, there are also benign tumors in some skin tumors (e.g. surface epithelial hyperplasia with severe dysplasia and mixed appendage tumor in skin).

5. Sinn et al. (16) obtained two separate transgenic mice strains, one carried the MMTV/Ha-ras oncogene while the other carried the MMTV/c-myc oncogene. These strains transmit their oncogenes to their progenies in a Mendelian fashion and show some benign proliferation of the lacrimal epithelium in the harderian gland. By interbreeding these two strains, Sinn et al. (16) found that many of the F_1 individuals developed more rapidly malignant tumors of mammary, salivary and lymphoid tissues but the tumors were developed stochastically and clonally. These results imply that while both the *ras* and the *myc* oncogenes are required for developing malignant tumors, at least one more event is needed for the development of tumors.

6. Franza et al. (10) showed by tissue cultures that the immortalization of cells and the ability of conferring responsiveness to the transforming effects of oncogenes like *ras* were separate properties of the oncogenes like *myc* or E1A (see also 29). This result suggests that besides immortalization and transformation, other events occur in converting normal stem cells to malignant tumors.

Recent studies on human colon and colorectal cancers have revealed that both oncogenes and antioncogenes are involved in these cancers. In fact, Bos et al. (5) reported frequent point mutations in codon 12 of the *ras* gene (mostly $c - ki - ras$) for colorectal cancers exclusively in somatic cells; further, most of these mutations were found in both the adenomatous (benign) and carcinomatous (malignant) regions of the tumor, suggesting that the mutation preceded the development of malignancy. A high frequency of point mutations involving a single base change at codon 12 of the $c - ki - ras$ gene was also found in many human colon tumor cell lines (9); the apparent lack of correlation between the presence or absence of mutant $c - ki - ras$ and the age, sex or race of the patient, or the anatomical localization of the tumors support the somatic nature of the *ras* gene mutational activation. On the other hand, Solomon et al. (17) and Bodmer et al. (4) reported an antioncogene at chromosome 5q21 for human colorectal carcinomas and for familial adenomatous polyposis; Vogelstein et al. (28) reported allelic deletions on the long arm of chromosome 18 (18q) and on the short arm of chromosome 17 (17p) for human colorectal cancers. These results imply that for human colorectal cancers, at least one dominantly acting oncogene (k-ras) and several tumor-suppressor genes are responsible. Further, as reported by Vogelstein et al. (28), the timing of ras-gene mutations and the allelic inactivation or deletion of 5q, 18q and 17p seem to correlate with the progression of colorectal cancers. It appears that ras-gene mutations are often relatively early events but are not usually the first genetic alternations; further, since inactivation of only one allele in 5q (referred to as the "Fap" gene) provides a selective growth advantage for cell proliferation while inactivation of both alleles only magnifies the proliferative effect, events other than the inactivation of the Fap allele are required for the development of colorectal cancer. These results suggest a multievent model as given in Section (6.2). Similar results have also been found in squamous cell carcinomas, large cell carcinoma and adenocarcinoma of the human lung which involve at least nine cancer genes, including the oncogenes myc, H-ras, k-ras, raf and jun and the antioncogenes from chromosomes 3p, 11p, 13q and 17p (30).

6.2 THE MODEL AND THE PROBABILITY GENERATING FUNCTIONS (PGF)

A schematic representation of the multi-event model is given in Figure 6.1. Like the multi-stage model of Armitage and Doll (1, 2), a cell may undergo a series of sequential changes or stages in becoming a transformed cell. In addition, a cell in each stage may not only progress to the next more malignant-like stage, but may proliferate (birth process) or die (death process).

6.2.1 Some Basic Assumptions

Let $I_0(t)$ and $I_j(t), j = 1, \ldots, k - 1$, be the numbers of normal stem cells and I_j cells, $j = 1, \ldots, k - 1$, at time t, respectively and let $T(t)$ be the

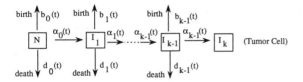

Figure 6.1 A multi-event model involving k stages.

number of tumors at time t. To derive mathematical theories, we make the following assumptions:

1.   At the time t_0 of birth, $I_0(t_0) = N_0$ is very large (say $N_0 \approx 10^6$–10^9). Since the mutation rate $\alpha_1(t)$ is usually very small and the cell proliferation rate of normal stem cells is usually nonnegative, this assumption implies that $I_0(t)$ is very large for all $t \geq t_0$ so that one may assume deterministic growth for normal stem cells; see Moolgavkar and Knudson (13).

2.   The probability that an I_{j-1} cell at time t yields one I_{j-1} cell and one I_j cell at time $t + \Delta t$ is $\alpha_j(t)\Delta t + o(\Delta t), j = 1,\ldots,k$, where $\lim_{\Delta t \to 0} o(\Delta t)/\Delta t = 0$.

3.   The I_j cells $(j = 1,\ldots,k-1)$ follow nonhomogeneous Feller-Arley birth-death processes with birth rate $b_j(t)$ and death rate $d_j(t)$ respectively for their proliferation and differentiation. If $b_j(t) - d_j(t) = \epsilon_j \exp(-\delta_j t)$ for some constants $\epsilon_j > 0$ and $\delta_j > 0$, the above birth-death process is then a stochastic Gompertz birth-death process (18).

4.   As in Moolgavkar and Knudson (13), it is assumed that the time period for the development of a clinically detectable cancer from a tumor cell is very short relative to the time for the transformation process. Thus, we assume that with probability one the tumor cells (I_k cells) develop into malignant tumors. Also, since the observed incidences for human beings were recorded using at least one year as time unit, for human cancers one may ignore the time elapsed between the generation of tumor cells and the development of tumors.

5.   All cells go through the above processes independently of other cells.

Given these assumptions, it immediately follows that the above process is Markov and that $I_j(t), j = 1,\ldots,k-1, T(t)$ form a continuous time multiple branching process with instantaneous transition probabilities during $[t, t + \Delta t]$ being given in assumptions (2)–(3) The Markov property follows from the fact that the future fate of a cell at time t depends only on the status of that cell at time t and is independent of past history.

6.2.2 The Probability Generating Functions (PGF)

Let $P_0(i_1,\ldots,i_k;t)$ be the conditional probability of $I_j(t) = i_j, j = 1,\ldots,k-1$ and $T(t) = I_k(t) = i_k$ given $I_0(t_0) = N_0$ and let $P_j(i_j,\ldots,i_k;t)$ be the conditional probability of $I_u(t) = i_u, u = j,\ldots,k$ given one I_j cell at time $t_0, j = 1,\ldots,k-1$.

Let $\psi(x_1\ldots,x_k;t_0,t) = \psi(t_0,t)$ be the *PGF* of $I_j(t), j = 1,\ldots,k-1$ and $T(t)$ given $I_0(t_0) = N_0$ and let $\phi_j(x_j,x_{j+1},\ldots,x_k;t_0,t) = \phi_j(t_0,t)$ be the PGF of $I_u(t), u = j,\ldots,k-1$ and $T(t)$ given one I_j cell at time $t_0, j = 1,\ldots,k-1$. Then

$$\psi(t_0,t) = \sum_{i_1}\cdots\sum_{i_k}[\prod_{u=1}^{k}x_u^{i_u}]P_0(i_1,\ldots,i_k;t)$$

$$\phi_j(t_0,t) = \sum_{i_j}\cdots\sum_{i_k}[\prod_{u=j}^{k}x_u^{i_u}]P_j(i_j,\ldots,i_k;t),$$

$j = 1,\ldots,k-1$.

To obtain $\phi_j(t_0,t)$, observe that for $j = 1,\ldots,k-1$, the Kolmogorov forward equation for $P_j(i_j,\ldots,i_k,t)$ is given by:

$$\frac{d}{dt}P_j(i_j,\ldots,i_k;t) = \sum_{u=j}^{k-1}\{(i_u-1)b_u(t)P_j(i_u-1,i_v,v \neq u,j \leq v \leq k;t)$$

$$+(i_u+1)d_u(t)P_j(i_u+1,i_v,v \neq u,j \leq v \leq k;t)$$

$$+i_u\alpha_u(t)P_j(i_{u+1}-1,i_v,v \neq u+1,j \leq v \leq k-1;t) - i_u[b_u(t)$$

$$+d_u(t) + \alpha_u(t)]P_j(i_j,\ldots,i_k;t)\}, \qquad j = 1,\ldots,k-1. \tag{6.2.1}$$

To obtain $\psi(t_0,t)$, observe that under assumption (1), it is expected that $I_0(t)$ is very large for all $t \geq t_0$ and that $I_0(t)\alpha_1(t)$ is finite for all $t \geq t_0$. Hence, one may assume that $I_0(t)$ is a deterministic function of t for $t \geq t_0$. Also, by the limiting theory of binomial distribution (law of small numbers), the number of mutations from normal stem cells to I_1 cells during $[t,t + \Delta t]$ would follow a Poisson distribution with parameter $I_0(t)\alpha_1(t)\Delta t + o(\Delta t)$.

THEOREM (6.2.1) Under conditions (2)–(5), $\phi_j(t_0,t), j = 1,\ldots,k - 1$, satisfy, respectively, the following first order partial differential equations:

For $j = 1, 2, \ldots, k - 1$,

$$\frac{\partial}{\partial t} \phi_j(t_0,t) = \sum_{u=j}^{k-1} \{(x_u - 1)[x_u b_u(t) - d_u(t)]$$

$$+ x_u(x_{u+1} - 1)\alpha_{u+1}(t)\} \frac{\partial}{\partial x_u} \phi_j(t_0,t), \phi_j(t_0,t_0) = x_j.$$

(6).2.2

Proof. Using Kolmogorov's forward equation, the proof of Theorem (6.2.1) is almost exactly the same as that of Theorem (3.2.2) in Chapter Three.

Solutions of equations (6.2.2) are in general very difficult to obtain; however, one may use these equations to approximate the incidence function of tumors and the moments of the number of tumors.

THEOREM (6.2.2) Suppose that $I_0(t)\alpha_1(t)$ is finite for all $t \geq t_0$ and that during $[t, t + \Delta t]$ the number of mutations from normal stem cells to I_1 cells follows a Poisson distribution with parameter $I_0(t)\alpha_1(t)\Delta t + o(\Delta t)$. If conditions (1)–(5) hold, then $\psi(t_0,t)$ is given by:

$$\psi(t_0,t) = \exp \{ \int_{t_0}^{t} I_0(u)\alpha_1(u)[\phi_1(u,t) - 1]d\,u \},$$

(6.2.3)

where $I_0(u)$ is the expected number of normal stem cells (I_0 cells) at time u given N_0 normal stem cells at $t = t_0$.

Proof. Partition the time interval $[t_0,t]$ by $L_j = [t_{j-1},t_j), j = 1,\ldots,n-1$ and $L_n = [t_{n-1},t_n]$, where $t = t_n$ and $t_j = t_0 + j\Delta t$, and let M_j be the number of mutations from normal stem cells to I_1 cells in $L_j, j = 1,\ldots,n$. Then, given the partition and given $(M_j, j = 1,\ldots,n)$, the conditional PGF of $I_u(t), u = 1,\ldots,k-1$ and $T(t)$ is

$$Q(t_0,t|\Delta t, M_j, j = 1,\ldots,n) = \prod_{j=1}^{n} [\phi_1(t_{j-1},t)]^{M_j}$$

(6.2.4)

Noting that the $M_j's$ are independent Poisson variates with parameters $I_0(t_{j-1})\alpha_1(t_{j-1})\Delta t + o(\Delta t)$ respectively, by taking the expectation over

$(M_j, j = 1, \ldots, n)$ we obtain from (6.2.4) the conditional *PGF* of $I_u(t), u = 1, \ldots, k - 1$ and $T(t)$ given the partition as,

$$\zeta_p(t_0, t; \Delta t) = \exp \left\{ \sum_{j=1}^{n} I_0(t_{j-1})\alpha_1(t_{j-1})[\phi_1(t_{j-1}, t) - 1]\Delta t + o(\Delta t) \right\}.$$

It follows that $\psi(t_0, t)$ is given by

$$\psi(t_0, t) = \lim_{\Delta t \to 0} \zeta_p(t_0, t; \Delta t) = \exp \left\{ \int_{t_0}^{t} I_0(x)\alpha_1(x)[\phi_1(x, t) - 1]dx \right\}.$$

If the normal stem cells follow a nonhomogeneous Feller-Arley birth-death process with birth rate $b_0(t)$ and death rate $d_0(t)$ for cell proliferation, then $I_0(t), I_j(t), j = 1, \ldots, k - 1$ and $T(t)$ follow a continuous time multiple branching process. In this case, the result of Theorem (6.2.2) can also be proved by using the results of continuous time multiple branching processes as given in Tan and Brown (20). In fact, we have the following theorem.

THEOREM (6.2.3) Suppose that the normal stem cells follow a nonhomogeneous Feller-Arley birth-death process with birth rate $b_0(t)$ and death rate $d_0(t)$ for cell proliferation. If N_0 is very large and $N_0\alpha_1(t)$ is finite for all $t \geq t_0$, then under conditions (2)–(5), to order $O(N_0^{-1})$, $\psi(t_0, t)$ is given by (6.2.3).

Proof. Noting the transition probabilities during $[t, t + \Delta t]$ as given in Table (6.1), the proof of Theorem (6.2.3) is almost exactly the same as that of Theorem (2) of Tan and Brown (20).

To use the results of Theorems (6.2.2) and (6.2.3), one would need the functions $I_0(t)$ and $\phi_1(u, t)$. The functional form of $I_0(t)$ depends on the growth pattern of different tissues. For the MVK two-stage models, Moolgavkar and Knudson (13) have assumed different growth patterns for different tissues to fit the theoretical incidence functions to observed incidence rates. For human cancers, the most common growth patterns are the exponential growth, logistic growth and Gompertz growth.

1. For breast cancer of women, Moolgavkar and Knudson (13) have shown that the growth of normal stem cells is best described by logistic

growth. In this case,

$$I_0(t) \cong N_0 \exp\left\{ \int_{t_0}^t \epsilon_N(x)dx \right\}\{1 - (N_0/M)$$

$$+ (N_0/M) \exp\left[\int_{t_0}^t \epsilon_N(x)dx\right]\}^{-1},$$

where M is the maximum number of normal stem cells, $\epsilon_N(x) = b_N(x) - d_N(x), b_N(t)$ and $d_N(t)$ is the birth rate and death rate at time t of normal stem cells (see 22).

2. For the lung and the colon, the growth pattern of normal stem cells is best described by Gompertz growth (12). Hence, one may take,

$$I_0(t) = N_0 \exp\{(\epsilon_N/\delta)[\exp(-\delta t_0) - \exp(-\delta t)]\},$$

where $\delta > 0$ and $\epsilon_N = b_0 - d_0$. (see 18).

3. For retinoblastoma, Wilm's tumor or Hodgkin's disease, the growth pattern of stem cells is fitted by exponential or Gamma growth curves (12, 13, 23). Hence, one may take,

$$I_0(t) = N_0 \exp[\epsilon_N(t - t_0)]$$

or

$$I_0(t) = a N_0(t - t_0)^{c-1} \exp[-\epsilon_N(t - t_0)]$$

where $\epsilon_0 = b_0 - d_0, c \ (c > 1)$ and $a \ (a > 0)$ are constants.

6.3 THE AGE-SPECIFIC INCIDENCE FUNCTION OF CANCER TUMORS

Let $\lambda(t)$ be the incidence rate of tumors at time t given N_0 normal stem cells at $t = t_0$. Denoting by $\psi'()$ the derivative of $\psi()$ with respect to t, by Theorem (2.1.1) of Chapter Two,

$$\lambda(t) = -\psi'(1,\ldots 1, 0; t_0, t)/\psi(1,\ldots, 1, 0; t_0, t), \tag{6.3.1}$$

where $\psi(1,\ldots, 1, 0; t_0, t) = [\psi(t_0,t)]_{(x_j=1,j=1,\ldots,k-1,x_k=0)}$.
To obtain $\lambda(t)$, let $\nu_1(t_0, t)$ and $u_{1k-1}(t_0, t)$ be defined by:

$$\nu_1(t_0, t) = -\{\partial\phi_1(t_0,t)/\partial t\}_{(x_j=1,j=1,\ldots k-1,x_k=0)}$$

and

$$u_{1k-1}(t_0, t) = \{\partial\phi_1(t_0,t)/\partial x_{k-1}\}_{(x_j=1,j=1,\ldots,k-1,x_k=0)}.$$

Then, from equation (6.2.2),

$$\nu_1(t_0,t) = \alpha_k(t)u_{1,k-1}(t_0,t)$$

and from (6.2.3),

$$\lambda(t) = \int_{t_0}^{t} I_0(x)\alpha_1(x)\nu_1(x,t)d\,u$$

$$= \alpha_k(t)\int_{t_0}^{t} I_0(x)\alpha_1(x)u_{1,k-1}(x,t)d\,x. \qquad (6.3.2)$$

Since in most cases, solution of equation (6.2.2) is extremely difficult to obtain, to compute $\lambda(t)$ by (6.3.2) is generally very difficult. For practical purposes, since $\alpha_k(t)$ is usually very small and since $\epsilon_j(t)$ is not big (≤ 0.1) for all $t \geq t_0, j = 1,\ldots,k-1$, one may approximate $\lambda(t)$ in (6.3.2) by approximating $u_{1k-1}(x,t)$ by $m_{k-1}(x,t)$, where $m_{k-1}(x,t) = \{\partial\phi_1(x,t)/\partial x_{k-1}\}_{(x_j=1,j=1,\ldots,k)}$.

To obtain $m_{k-1}(x,t)$, let $m_j(x,t)$ be defined by:

$$m_j(x,t) = \{\partial\phi_1(x,t)/\partial x_j\}_{(x_v=1,v=1,\ldots,k)}, j = 1,2,\ldots,k.$$

Then, by taking the derivative with respect to x_j and putting $x_v = 1, v = 1,\ldots,k$, we obtain from equation (6.2.2):

$$\frac{d}{dt}m_1(t_0,t) = \epsilon_1(t)m_1(t_0,t), m_1(t_0,t_0) = 1; \qquad (6.3.3)$$

for $j = 2,\ldots,k-1$,

$$\frac{d}{dt}m_j(t_0,t) = \alpha_j(t)m_{j-1}(t_0,t) + \epsilon_j m_j(t_0,t), m_j(t_0,t_0) = 0, \qquad (6.3.4)$$

and

$$\frac{d}{dt}m_k(t_0,t) = \alpha_k(t)m_{k-1}(t_0,t), m_k(t_0,t_0) = 0, \qquad (6.3.5)$$

where $\epsilon_j(t) = b_j(t) - d_j(t)$.

From (6.3.3), we obtain:

$$m_1(t_0,t) = \exp\left\{\int_{t_0}^{t} \epsilon_1(x)d\,x\right\}; \qquad (6.3.6)$$

from (6.3.4), we have, for $j = 2,\ldots,k-1$:

$$m_j(t_0,t) = \int_{t_0}^{t} m_{j-1}(t_0,x)\alpha_j(x)\exp[\int_{x}^{t} \epsilon_j(y)d\,y]d\,x \qquad (6.3.7)$$

and from (6.3.5),

$$m_k(t_0, t) = \int_{t_0}^{t} \alpha_k(x) m_{k-1}(t_0, x) dx. \tag{6.3.8}$$

From (6.3.6) and (6.3.7), one has for $t_0 \leq x < t$:

$$m_{k-1}(x, t) = \int_x^t m_{k-2}(x, x_1) \alpha_{k-1}(x_1) \exp[\int_{x_1}^t \epsilon_{k-1}(y) dy] dx_1$$

$$= \ldots = \int_x^t \int_x^{x_1} \int_x^{x_2} \ldots \int_x^{x_{k-3}} \int_x^{x_{k-2}} \exp[\int_x^{x_{k-2}} \epsilon_1(y) dy] \tag{6.3.9}$$

$$\times \prod_{u=2}^{k-1} \{\alpha_u(x_{k-u}) \exp[\int_{x_{k-u}}^{x_{k-u-1}} \epsilon_u(y) dy] dx_{k-u}\}.$$

Replacing $u_{1,k-1}(x, t)$ by $m_{k-1}(x, t)$ of (6.3.9) in $\lambda(t)$, we obtain with $t = x_0$:

$$\lambda(t) \cong \alpha_k(t) \int_{t_0}^{t} I_0(x) \alpha_1(x) m_{k-1}(x, t) dx$$

$$= \alpha_k(t) \int_{t_0}^t \int_x^t \int_x^{x_1} \ldots \int_x^{x_{k-3}} I_0(x) \tag{6.3.10}$$

$$\times \prod_{u=1}^{k-1} \{\alpha_u(x_{k-u}) \exp[\int_{x_{k-u}}^{x_{k-u-1}} \epsilon_u(y) dy] dx_{k-u}\} dx$$

6.4 THE EXPECTED VALUE AND THE VARIANCE OF THE NUMBER OF CANCER TUMORS

Using (6.2.3), theoretically one may derive the probability distribution of the number of tumors. In practice, however, except for $k \leq 2$, it is rather difficult to compute the probabilities. However, using equation (6.2.2) and formula (6.2.3) one may readily evaluate the moments of the number of tumors. In particular, one may readily evaluate the expected value $\mu(t_0, t)$ and the variance $V(t_0, t)$ of the number of tumors given $I_0(t_0) = N_0$.

To illustrate, we put:

$$C_{u,v}(s, t) = C_{uv}(s, t) = \text{Covariance of } (I_u(t), I_v(t))$$

$$\text{given} \quad I_1(s) = 1, u, v = 1, \ldots, k.$$

Then,

$$\nu_{u,v}(s,t) = \{\partial^2 \phi_1(s,t)/\partial x_u \partial x_v\}_{(x_j=1,j=1,\dots,k)} = C_{uv}(s,t)$$
$$-\delta_{uv}m_u(s,t) + m_u(s,t)m_v(s,t), \delta_{uv} \quad \text{being the Kronecker's } \delta. \tag{6.4.1}$$

Thus, by taking derivatives of (6.2.3) and putting $x_v = 1$ for $v = 1,\dots,k$, we obtain the expected value $\mu(t_0,t)$ and the variance $V(t_0,t)$ of the number of tumors given $I_0(t_0) = N_0$ as

$$\mu(t_0,t) = \int_{t_0}^{t} I_0(u)\alpha_1(u)m_k(u,t)d\,u, \tag{6.4.2}$$

and

$$V(t_0,t) = \int_{t_0}^{t} I_0(u)\alpha_1(u)[C_{kk}(u,t) + m_k^2(u,t)]d\,u. \tag{6.4.3}$$

Note that $m_k(u,t)$ is available from (6.3.8).

To obtain $C_{kk}(s,t)$, observe that from (6.4.1),

$$\frac{\partial}{\partial t}\nu_{uv}(s,t) + \delta_{uv}\frac{\partial}{\partial t}m_u(s,t) - m_u(s,t)\frac{\partial}{\partial t}m_v(s,t)$$

$$- m_v(s,t)\frac{\partial}{\partial t}m_u(s,t) = \frac{d}{dt}C_{uv}(s,t). \tag{6.4.4}$$

Taking derivatives of both sides of equation (6.2.2), setting $x_j = 1, j = 1,\dots,k$ and noting (6.4.4), we obtain, with $m_0(s,t) = C_{0j}(s,t) = 0$ for all $j = 1,\dots,k$:

$$\frac{d}{dt}C_{ij}(s,t) = \delta_{ij}\{[b_i(t) + d_i(t)]m_i(s,t) + \alpha_i(t)m_{i-1}(s,t)\}$$

$$+ [\alpha_i(t)C_{i-1j}(s,t) + \alpha_j(t)C_{ij-1}(s,t)]$$

$$+ [\epsilon_i(t) + \epsilon_j(t)]C_{ij}(s,t), \quad C_{ij}(s,s) = 0, 1 \le i,j < k; \tag{6.4.5}$$

$$\frac{d}{dt}C_{ik}(s,t) = \alpha_i(t)C_{i-1k}(s,t) + \alpha_k(t)C_{i,k-1}(s,t)$$

$$+ \epsilon_i(t)C_{ik}(s,t), C_{ik}(s,s) = 0,$$

$$1 \le i < k; \tag{6.4.6}$$

and

$$\frac{d}{dt}C_{kk}(s,t) = 2\alpha_k(t)C_{k-1,k}(s,t) + \alpha_k(t)m_{k-1}(s,t), C_{kk}(s,s)$$

$$= 0. \tag{6.4.7}$$

From (6.4.7),

$$C_{kk}(s,t) = \int_s^t \alpha_k(x)[2C_{k-1,k}(s,x) + m_{k-1}(s,x)]dx. \tag{6.4.8}$$

In (6.4.8), $m_{k-1}(s,x)$ is available from (6.3.7) while $C_{k-1,k}(s,x)$ can readily be obtained by solving the simultaneous equations (6.4.5) and (6.4.6). To illustrate, let $k = 3$. Then equations (6.4.5), (6.4.6) and (6.4.7) reduce to:

$$\frac{d}{dt}C_{11}(s,t) = [b_1(t) + d_1(t)]m_1(s,t) + 2\epsilon_1(t)C_{11}(s,t),$$

$$\frac{d}{dt}C_{12}(s,t) = \alpha_2(t)C_{11}(s,t) + [\epsilon_1(t) + \epsilon_2(t)]\,C_{12}(s,t),$$

$$\frac{d}{dt}C_{22}(s,t) = [b_2(t) + d_2(t)]m_2(s,t) + \alpha_2(t)[2C_{12}(s,t) + m_1(s,t)]$$

$$+2\epsilon_2(t)C_{22}(s,t), \ C_{ij}(s,s) = 0, 1 \leq i,j \leq 2; \tag{6.4.9}$$

$$\frac{d}{dt}C_{13}(s,t) = \alpha_3(t)C_{12}(s,t) + \epsilon_1(t)C_{13}(s,t),$$

$$\frac{d}{dt}C_{23}(s,t) = \alpha_2(t)C_{13}(s,t) + \alpha_3(t)C_{22}(s,t) + \epsilon_2(t)C_{23}(s,t), C_{i3}(s,s)$$

$$= 0, i = 1,2; \tag{6.4.10}$$

$$\frac{d}{dt}C_{33}(s,t) = \alpha_3(t)[2C_{23}(s,t) + m_2(s,t)], \ C_{33}(s,s) = 0. \tag{6.4.11}$$

From (6.4.9),

$$C_{11}(s,t) = \int_s^t [b_1(u) + d_1(u)]m_1(s,u) \exp[2\int_u^t \epsilon_1(y)dy]du,$$

$$C_{12}(s,t) = \int_s^t \alpha_2(u)C_{11}(s,u) \exp\{\int_u^t [\epsilon_1(y) + \epsilon_2(y)]dy\}du,$$

and

$$C_{22}(s,t) = \int_s^t \{[b_2(u) + d_2(u)]m_2(s,u) + \alpha_2(u)[2C_{12}(s,u)$$

$$+ m_1(s,u)]\}\exp[2\int_u^t \epsilon_2(y)dy]du; \tag{6.4.12}$$

from (6.4.10),

$$C_{13}(s,t) = \int_s^t \alpha_3(u)C_{12}(s,u) \exp[\int_u^t \epsilon_1(y)dy]du,$$

$$C_{23}(s,t) = \int_s^t [\alpha_2(u)C_{13}(s,u) + \alpha_3(u)C_{22}(s,u)] \exp[\int_u^t \epsilon_2(y)dy]du;$$

and from (6.4.11),

$$C_{33}(s,t) = \int_s^t \alpha_3(u)[2C_{23}(s,u) + m_2(s,u)]du.$$

6.5   THE HOMOGENEOUS MULTIEVENT MODEL

If $\alpha_j(t) = \alpha_j, b_j(t) = b_j, d_j(t) = d_j$ for all $j = 1,\ldots,k-1$, and $\alpha_k(t) = \alpha_k$, then the model given in Section (6.2) is time homogeneous. In this case $\phi_j(t_0,t) = \phi_j(t-t_0), j = 1,\ldots,k-1$ so that we assume $t_0 = 0$. Furthermore, as shown in Section (3.3) of Chapter Three, $\phi_{k-1}(t)$ satisfies the following Ricatti equation:

$$\frac{d}{dt}\phi_{k-1}(t) = b_{k-1}\phi_{k-1}^2(t) + [\alpha_k x_k - (b_{k-1} + d_{k-1} + \alpha_k)]\phi_{k-1}(t)$$

$$+d_{k-1}, \quad \phi_{k-1}(0) = x_{k-1}, \tag{6.5.1}$$

Similarly, it can easily be shown by the same approach as given in Section (3.3.1) of Chapter Three that $\phi_j(t)$, $j = 1,\ldots,k-2$, satisfies, respectively, the following Ricatti equations:

$$\frac{d}{dt}\phi_j(t) = b_j\phi_j^2(t) + [\alpha_{j+1}\phi_{j+1}(t) - (b_j + d_j + \alpha_{j+1})]\phi_j(t)$$

$$+d_j, \quad \phi_j(0) = x_j, \quad j = 1,2,\ldots,k-2. \tag{6.5.2}$$

Let $y_2 > y_1$ be defined by:

$$2b_{k-1}y_2 = (b_{k-1} + d_{k-1} + \alpha_k - x_k\alpha_k) + h_k(x_k)$$

and

$$2b_{k-1}y_1 = (b_{k-1} + d_{k-1} + \alpha_k - x_k\alpha_k) - h_k(x_k),$$

where

$$h_k(x_k) = [(b_{k-1} + d_{k-1} + \alpha_k - x_k\alpha_k)^2 - 4b_{k-1}d_{k-1}]^{\frac{1}{2}}.$$

Then, by formula (3.3.3) of Chapter Three, the solution of (6.5.1) is

$$\phi_{k-1}(t) = \{y_2(x_{k-1} - y_1) + y_1(y_2 - x_{k-1})\exp[(y_2 - y_1)b_{k-1}t]\}$$

$$\times\{(x_{k-1} - y_1) + (y_2 - x_{k-1})\exp[(y_2 - y_1)b_{k-1}t]\}^{-1}. \tag{6.5.3}$$

Since equation (6.5.2) is a nonlinear Ricatti equation, the solution of equation (6.5.2) is very complicated in general cases. If $d_j[1 - \phi_j(t)] = 0$, then equation (6.5.2) reduces to:

$$\frac{d}{dt}\phi_j(t) = b_j\phi_j^2(t) + [\alpha_{j+1}\phi_{j+1}(t) - (b_j + \alpha_{j+1})]\phi_j(t),$$

$$\phi_j(0) = x_j. \quad (6.5.4)$$

The solution of equation (6.5.4) is given by:

$$\phi_j(t) = x_j \exp[\alpha_{j+1}\int_0^t \phi_{j+1}(u)d\,u - (b_j + \alpha_{j+1})t]$$

$$\times\{1 - b_jx_j\int_0^t \exp[\alpha_{j+1}\int_0^u \phi_{j+1}(x)d\,x - (b_j + \alpha_{j+1})u]d\,u\}^{-1}, \quad (6.5.5)$$

$$j = 1,2,\ldots,k - 2.$$

(For details, see Section (4.6.1) of Chapter Four). Hence, if $d_j = 0$, the solution of equation (6.5.2) is given by (6.5.5).

Alternatively, if $\alpha_{j+1} \cong 0$, then equation (6.5.2) approximately reduces to:

$$\frac{d}{dt}\phi_j(t) = b_j\phi_j^2(t) - (b_j + d_j)\phi_j(t) + d_j$$

$$= [b_j\phi_j(t) - d_j][\phi_j(t) - 1], \phi_j(0) = x_j. \quad (6.5.6)$$

Using the approach of Section (2.5.2) of Chapter Two, the solution of (6.5.6) is

$$\phi_j(t) = [(1 - x_j)d_j + (x_jb_j - d_j)\exp(-\epsilon_jt)]$$

$$\times[(1 - x_j)b_j + (x_jb_j - d_j)\exp(-\epsilon_jt)]^{-1}, \quad \text{where} \quad \epsilon_j = b_j - d_j. \quad (6.5.7)$$

Note that (6.5.7) is the PGF of a homogeneous Feller-Arley birth-death process with birth rate b_j and death rate d_j. Note also that $\phi_j(t)$ in (6.5.7) is $\phi_j(t) = 1$ if $x_j = 1$. Thus, if $\alpha_{j+1} \approx 0$ and if $x_j = 1$, then the solution of equation (6.5.2) is approximately $\phi_j(t) \approx 1$. This result leads further to $\phi_u(t) \approx 1$ for $1 \leq u \leq j$ when $\alpha_{j+1} \approx 0$ and when $x_v = 1, v = 1,\ldots,j$. These results will be used to approximate the incidence function in the next section.

6.5.1   The Age-Specific Incidence Function of Cancer Tumors

From (6.3.2), the incidence function $\lambda(t)$ is given by:

$$\lambda(t) = \alpha_1 \alpha_k \int_0^t I_0(x) u_{1,k-1}(t-x) dx, \qquad (6.5.8)$$

where

$$u_{j,k-1}(t) = [d\phi_j(t)/dx_{k-1}]_{(x_v=1, v=j,...,k-1, x_k=0)}.$$

Let $\phi_j(1,\ldots,1,0;s,t) = \phi_j(1,\ldots,1,0;t-s) = \phi_{10(j)}(t-s)$.
From equation (6.5.2), for $1 \le j \le k-2$,

$$\frac{d}{dt} u_{j,k-1}(t) = 2b_j \phi_{10(j)}(t) u_{j,k-1}(t) + \alpha_{j+1} u_{j+1,k-1}(t) \phi_{10(j)}(t)$$

$$+ [\alpha_{j+1} \phi_{10(j+1)}(t) - (b_j + d_j + \alpha_{j+1})] u_{j,k-1}(t),$$

$$u_{j,k-1}(0) = 0; \qquad (6.5.9)$$

from (6.5.1),

$$\frac{d}{dt} u_{k-1,k-1}(t) = 2b_{k-1} \phi_{10(k-1)}(t) u_{k-1,k-1}(t)$$

$$-(b_{k-1} + d_{k-1} + \alpha_k) u_{k-1,k-1}(t), \qquad u_{k-1,k-1}(0) = 1. \qquad (6.5.10)$$

From (6.5.10), one has:

$$u_{k-1,k-1}(t)$$

$$= \exp\{-(b_{k-1} + d_{k-1} + \alpha_k)t + 2b_{k-1} \int_0^t \phi_{10(k-1)}(x) dx\}; \quad (6.5.11)$$

from (6.5.9), for $1 \le j \le k-2$,

$$u_{j,k-1}(t) = \alpha_{j+1} \int_0^t u_{j+1,k-1}(x) \phi_{10(j)}(x) \exp[\int_x^t a_j(y) dy] dx, \quad (6.5.12)$$

where $a_j(t) = 2b_j \phi_{10(j)}(t) + \alpha_{j+1} \phi_{10(j+1)}(t) - (b_j + d_j + \alpha_{j+1})$.

On substituting $u_{1,k-1}(t)$ of (6.5.11) into $\lambda(t)$ given in (6.5.8), one may evaluate $\lambda(t)$ for different values of t. However, to compute $u_{i,k-1}(t)$ from (6.5.12), one would need $\phi_{10(j)}(t)$ for $1 \le j \le k-1$ which in turn requires the solution $\phi_j(t)$ of equations (6.5.2). Note that $\phi_{k-1}(t)$ is available from (6.5.3) so that $\phi_{10(k-1)}(t)$ is available. To obtain $\phi_{10(j)}(t)$ for $j = 1,\ldots,k-2$, we consider two special cases:

(a) **Case 1:** $d_j = 0$ for all $j = 1,\ldots,k-2$. In this case, $\phi_j(t)$ and hence $\phi_{10(j)}(t)$ for $j = 1,2,\ldots,k-2$ is available from (6.5.5).

(b) Case 2: $\alpha_{k-1} \approx 0$.

If $\alpha_{k-1} \approx 0$, then by result (6.5.7), approximately $\phi_{10(j)}(t) \approx 1$ for all $j = 1,\ldots,k-2$. In this case, if $\epsilon_j = b_j - d_j, j = 1,\ldots,k-2$, is small ($\leq 0.1$), $u_{j,k-1}(t)$ in (6.5.12) is then approximately given by:

$$u_{j,k-1}(t) \cong \alpha_{j+1} \int_0^t u_{j+1,k-1}(x)\exp[\epsilon_j(t-x)]dx \quad j = 1,2,\ldots,k-2.$$

It follows that

$$u_{1k-1}(t) \cong \alpha_2 \int_0^t \exp[\epsilon_1(t-x)]u_{2,k-1}(x)dx$$

$$= \alpha_2\alpha_3 \int_0^t \int_0^{x_1} \exp[\epsilon_1(t-x_1) + \epsilon_2(x_1-x_2)]u_{3,k-1}(x_2)dx_2 dx_1$$

$$= \alpha_2\alpha_2\ldots\alpha_{k-1} \int_0^t \int_0^{x_1} \ldots \int_0^{x_{k-3}} \exp[\epsilon_1(t-x_1) \quad (6.5.13)$$

$$+ \sum_{u=2}^{k-2} \epsilon_u(x_{u-1} - x_u)] u_{k-1k-1}(x_{k-2})dx_{k-2}dx_{k-3}\ldots dx_1.$$

On substituting (6.5.13) into (6.5.8),

$$\lambda(t) \cong \alpha_1\alpha_2\ldots\alpha_k \int_0^t I_0(x_1)\left\{ \int_0^{t-x_1} \int_0^{x_2} \ldots \int_0^{x_{k-2}} \exp[\epsilon_1(t-x_1-x_2) \right.$$

$$+ \sum_{u=2}^{k-2} \epsilon_u(x_u - x_{u+1})]u_{k-1,k-1}(x_{k-1})dx_{k-1}dx_{k-2}\ldots dx_2 \bigg\}dx_1. \quad (6.5.14)$$

Note that if one replaces the condition $\alpha_{k-1} \approx 0$ by $\alpha_k \approx 0$, then $\phi_{10(j)}(t) \approx 1$ for all $j = 1,\ldots,k-1$. Since $\phi_{10(k-1)}(t) \approx 1$ implies that $u_{k-1,k-1}(t) \approx \exp(\epsilon_{k-1}t)$, if $\epsilon_j = b_j - d_j, j = 1,\ldots,k-1$, is small, then (6.5.14) reduces to

$$\lambda(t) \cong \alpha_1\alpha_2\ldots\alpha_k \int_0^t I_0(x_1)\left\{ \int_0^{t-x_1} \int_0^{x_2} \ldots \int_0^{x_{k-2}} \exp[\epsilon_1(t-x_1-x_2) \right.$$

$$+ \sum_{u=2}^{k-2} \epsilon_u(x_u - x_{u+1}) + \epsilon_{k-1}x_{k-1}]dx_{k-1}dx_{k-2}\ldots dx_2 \bigg\}dx_1. \quad (6.5.15)$$

We now proceed to show that approximation (6.5.15) is equivalent to approximating $\lambda(t)$ by replacing $u_{1k-1}(x,t)$ by $m_{k-1}(x,t)$. That is, approximation (6.5.15) is exactly the same as the approximation using (6.3.10). To see this, observe that from (6.3.9), the expected number of

I_{k-1} cells at time t given one I_1 cell at time y is, with $y_0 = t$,

$$m_{k-1}(y,t) = \alpha_2 \ldots \alpha_{k-1} \int_y^t \int_y^{y_1} \int_y^{y_2} \ldots \int_y^{y_{k-3}} \exp[\epsilon_1(y_{k-2} - y)$$

$$+ \sum_{u=2}^{k-1} \epsilon_u(y_{k-u-1} - y_{k-u})] dy_{k-2} dy_{k-3} \ldots dy_1. \qquad (6.5.16)$$

Note that in (6.5.16), $y \leq y_{k-2} \leq \ldots \leq y_2 \leq y_1 \leq t = y_0$. Interchanging the order of integration, $m_{k-1}(y,t)$ in (6.5.16) can be written as,

$$m_{k-1}(y,t) = \alpha_2 \ldots \alpha_{k-1} \int_y^t \int_{y_{k-2}}^t \ldots \int_{y_2}^t \exp[\epsilon_1(y_{k-2} - y)$$

$$+ \sum_{u=2}^{k-1} \epsilon_u(y_{k-u-1} - y_{k-u})] dy_1 dy_2 \ldots ky_{k-3} dy_{k-2}. \qquad (6.5.17)$$

Let $x_j = t - y_{k-j}, j = 2, \ldots, k-1$. Then, $x_{k-1} = t - y_1 = y_0 - y_1$ and for $u = 2, \ldots, k-2, y_{k-u-1} - y_{k-u} = x_u - x_{u+1}, y_{k-2} - y = t - y - x_2$ and $0 \leq x_{k-1} \leq x_{k-2} \leq \ldots \leq x_2 \leq t - y$. Making the transformation $x_j = t - y_{k-j}, j = 2, \ldots, k-1$, then $m_{k-1}(y,t)$ in (6.5.17) becomes, with $x_k = 0$,

$$m_{k-1}(y,t) = \alpha_2 \ldots \alpha_{k-1} \int_0^{t-y} \int_0^{x_2} \ldots \int_0^{x_{k-2}} \exp[\epsilon_1(t - y - x_2)$$

$$+ \sum_{u=2}^{k-1} \epsilon_u(x_u - x_{u+1})] dx_{k-1} dx_{k-2} \ldots dx_2. \qquad (6.5.18)$$

Or, equivalently, $u_{1,k-1}(x_1,t) \cong m_{k-1}(x_1,t)$ as given in (6.3.9) of Section (6.3) so that the $\lambda(t)$ in (6.5.15) is exactly the same $\lambda(t)$ as given in (6.3.10).

6.5.2 The Armitage–Doll Multistage Model

In this model, $\epsilon_j = 0$ for all $j = 0, 1, 2, \ldots, k-1$, hence, $I_0(x) = N_0$ and (6.5.18) reduces to

$$u_{1,k-1}(t) \cong \alpha_2 \alpha_3 \ldots \alpha_{k-1} \int_0^t \int_0^{x_1} \ldots \int_0^{x_{k-3}} dx_{k-2} dx_{k-3} \ldots dx_1$$

$$= \alpha_2 \ldots \alpha_{k-1} \{ t^{k-2}/(k-2)! \}. \qquad (6.5.19)$$

It follows that

$$\lambda(t) \cong \alpha_1\alpha_2\ldots\alpha_k[(k-2)!]^{-1}\int_0^t N(x)(t-x)^{k-2}dx$$

$$= N_0\alpha_1\alpha_2\ldots\alpha_k[t^{k-1}/(k-1)!], \qquad (6.5.20)$$

(6.5.20) is the incidence function of the Armitage–Doll multistage model in which the cell differentiation of normal stem cells and intermediate cells are ignored.

6.6 THE NONHOMOGENEOUS PIECEWISE MULTIEVENT MODEL

In many practical situations, the time interval $[t_0, t]$ can be partitioned into nonoverlapping intervals in each of which one may assume a homogeneous model. Examples of these models include animal experiments involving initiators and promoters, and breast cancer in women in which menarche, first pregnancy and menopause form natural partition points. In this section, we thus consider a situation in which $[t_0, t]$ is partitioned into m nonoverlapping intervals $L_j = [t_{j-1}, t_j), j = 1, \ldots, m-1$ and $L_m = [t_{m-1}, t_m]$ with $t = t_m$ and assume that in $L_j, \alpha_i(t) = \alpha_{ij} = \alpha_{ij}, b_i(t) = b_{ij} = b_{ij}, d_i(t) = d_{ij} = d_{ij}, i = 1, \ldots, k-1, \alpha_k(t) = \alpha_{kj} = \alpha_{kj}, j = 1, \ldots, m$.

Given these specifications for the model, then in L_m, the model is a homogeneous model given in Section (6.5) with $\alpha_i, b_i, d_i, i = 1, \ldots, k-1, \alpha_k$ being replaced by $\alpha_{im}, b_{im}, d_{im}, i = 1, \ldots, k-1$ and α_{km}, respectively. Hence, if $t_{m-1} \leq s < t \leq t_m$ and if in (6.5.1) and (6.5.2) one replaces $(\alpha_i, b_i, d_i, i = 1, \ldots k-1)$ and α_k by $(\alpha_{im}, b_{im}, d_{im}, i = 1, \ldots, k-1)$ and α_{km} respectively, $\phi_{k-1}(s,t) = \phi_{k-1}(t-s)$ satisfies equation (6.5.1) and $\phi_i(s,t), i = 1, \ldots, k-2$, satisfies equations (6.5.2) respectively. To emphasize the dependence of $\phi_i(s,t)$ and $\phi_{k-1}(s,t)$ on t_m, in what follows we write $\phi_i(s,t)$ as $\phi_{i,m}(s,t), i = 1, \ldots, k-2$ and write $\phi_{k-1}(s,t)$ as $\phi_{k-1,m}(s,t)$ respectively, unless otherwise stated.

6.6.1 The Probability Generating Functions

To obtain $\psi(t_0, t)$ and $\phi_i(t_0, t), i = 1, \ldots, k-1$, consider the following set of differential equations:

$$\frac{\partial}{\partial t}\eta_i(s,t) = \sum_{u=i}^{k-1}[(x_u-1)(x_u b_{uj} - d_{uj})]$$

$$+x_u(x_{u+1} - 1)\alpha_{u+1j}] \frac{\partial}{\partial x_u} \eta_i(s,t), t_{j-1} \leq s < t_j \leq t,$$

$$\eta_i(s,s) = x_i, i = 1,\ldots,k-1, j = 1,\ldots,m. \tag{6.6.1}$$

For $t_{j-1} \leq s < t_j, j = 1, \ldots, m$, with $t = t_m$, let $\zeta_{i,j}(s,t) = \zeta_{i,j}(x_i,\ldots,x_k; s,t)$ be the solution of equation (6.6.1) , $i = 1,\ldots,k-1, j = 1,\ldots,m$. By the results of Section (6.5), if $t_{j-1} \leq s < t_j, j = 1,\ldots,m$ and if one replaces $(b_i,d_i,\alpha_i), i = 1,\ldots,k-1$ and α_k by $(b_{ij},d_{ij},\alpha_{ij}), i = 1,\ldots,k-1$ and α_{kj} respectively, $\zeta_{k-1j}(s,t)$ also satisfies equations (6.5.1) and $\zeta_{i,j}(s,t), i = 1,\ldots,k-2$, also satisfies equations (6.5.2).

To obtain $\psi(t_0,t)$ and $\zeta_{i,j}(s,t), i = 1,\ldots,k-1$, let for $t_{m-1} \leq s < t \leq t_m$ and $i = 1,\ldots,k-1$,

$$g_{i,m}(s,t) = \zeta_{i,m}(s,t) = \zeta_{i,m}(x_i,\ldots,x_k; s,t), \tag{6.6.2}$$

and define for $t_{j-1} \leq s < t_j, j = 1,\ldots,m-1$ and $i = 1,\ldots,k-1$,

$$g_{i,j}(s,t) = \zeta_{i,j}[g_{i,j+1}(t_j,t),\ldots,g_{k-1j+1}(t_j,t),x_k; s,t_j]. \tag{6.6.3}$$

Then by using the chain rule of differentiation and by mathematical induction, it can be shown that $g_{i,j}(s,t), i = 1,\ldots,k-1, j = 1,\ldots,m$ satisfies the equations (6.2.2) with $s = t_0$. (The proof, although more complicated, is almost exactly the same as that of proving formula (4.6.6) in Section (4.6.1) of Chapter Four.) Hence, as in the previous three chapters, $\psi(t_0,t)$ in (6.2.2) becomes:

$$\psi(t_0,t) \cong exp\{\int_{t_0}^{t} I_0(u)\alpha_1(u)[\phi_1(u,t) - 1]du\}$$

$$= exp\{\sum_{j=1}^{m} \int_{t_{j-1}}^{t_j} I_0(u)\alpha_1(u)[\zeta_1(u,t) - 1]du\} \tag{6.6.4}$$

$$= exp\{\sum_{j=1}^{m} N_{j-1} \int_{t_{j-1}}^{t_j} X_{j-1}(u)\alpha_1(u)[g_{1j}(u,t) - 1]du\}$$

where $t = t_m, N_{j-1}$ is the expected number of normal stem cells at time t_{j-1} and $X_{j-1}(u)$ the expected number of normal stem cells at time $u(\geq t_{j-1})$ given one normal stem cell at time t_{j-1}.

Note that formula (6.6.4) is an extension of a result of Tan and Gastardo (19) of the multievent models.

6.6.2 The Age-Specific Incidence Function of Cancer Tumors

By (6.3.2) with $t = t_m$, one has:

$$\lambda(t) = \alpha_k(t) \int_{t_0}^{t} I_0(x)\alpha_1(x)u_{1,k-1}(x,t)dx$$

$$= \alpha_{km} \sum_{j=1}^{m} \int_{t_{j-1}}^{t_j} I_0(x)\alpha_1(x)u_{1,k-1}(x,t)dx$$

$$= \alpha_{km} \sum_{j=1}^{m} N_{j-1}\alpha_{1j} \int_{t_{j-1}}^{t_j} X_{j-1}(x)u_{1,k-1}(x,t)dx.$$

If $\alpha_k(t)$ is very small ($\alpha_k(t) \approx 0$) and if $\epsilon_j(t), j = 1,\ldots,k-1$, is small ($\leq$ 0.1), then as shown in Section (6.3), $u_{1,k-1}(x,t)$ is closely approximated by the expected number $m_{k-1}(x,t)$ of I_{k-1} cells at time t given one I_1 cell at time x ($t \geq x$). Hence, if $\alpha_k(t) \approx 0$ and if $\epsilon_j(t), j = 1,\ldots,k-1$, is small,

$$\lambda(t) \approx \alpha_{km} \sum_{j=1}^{m} N_{j-1}\alpha_{1j} \int_{t_{j-1}}^{t_j} X_{j-1}(x)m_{k-1}(x,t)dx. \tag{6.6.5}$$

To obtain $m_{k-1}(x,t)$ and $\lambda(t)$, let $m_{i,u}(s,t)$ be the expected number of I_u cells at time t given one $I_i(u \geq i)$ cell at time s so that $m_{k-1}(s,t) = m_{1,k-1}(s,t)$.

For $t_{j-1} \leq s < t \leq t_j, j = 1,\ldots,m$ and $u \geq i, i,u = 1,\ldots,k-1$, define $\mu_{i,u(j)}(s,t)$ by:

$$\mu_{i,u(j)}(s,t) = [\partial \zeta_{i,j}(s,t)/\partial x_u]_{(x_v = 1, v = i,\ldots,k)}.$$

Then by (6.6.2), and (6.6.3) and by applying the chain rule of differentiation, we have, for $t_{m-1} \leq s < t \leq t_m$ and $i \leq u, i,u = 1,\ldots,k-1$:

$$m_{i,u}(s,t) = \mu_{i,u(m)}(s,t); \tag{6.6.6}$$

and for $t_{j-1} \leq s < t_j < t, j = 1,\ldots,m-1$ and $i \leq u, i,u = 1,\ldots,k-1$,

$$m_{i,u}(s,t) = \mu_{i,i(j)}(s,t_j)m_{i,u}(t_j,t) + \mu_{i,i+1(j)}(s,t_j)m_{i+1,u}(t_j,t)$$
$$+ \ldots + \mu_{i,u(j)}(s,t_j)m_{u,u}(t_j,t) \tag{6.6.7}$$

where $m_{ij}(t,t) = \delta_{ij}$, the Kronecker's δ.

From formulas (6.6.6) and (6.6.7), $m_{k-1}(s,t)$ for all $t_{j-1} \leq s < t_j \leq t, j = 1,\ldots,m$, can readily be computed if for all $t_{j-1} \leq s < t_j \leq t, j = 1,\ldots,m$ and for all $i \leq u, i,u = 1,\ldots,k-1$, the $\mu_{i,u(j)}(s,t)'s$ are available.

To obtain $\mu_{i,u(j)}(s,t_j)$, observe that as we have shown in the previous section, $\zeta_{i,j}(s,t) = \zeta_{i,j}(x_i,\ldots,x_k;s,t)$ with $t_{j-1} \leq s < t \leq t_j, i = 1,\ldots,k-1, j = 1,\ldots,m$, satisfy, respectively, the following set of Ricatti equations:

$$\frac{d}{dt}\zeta_{i,j}(s,t) = b_{ij}\zeta_{i,j}^2(s,t) + [\alpha_{i+1j}\zeta_{i+1,j}(s,t)$$
$$-(b_{ij} + d_{ij} + \alpha_{i+1j})\zeta_{i,j}(s,t) + d_{ij}, \tag{6.6.8}$$

where $\zeta_{k,j}(s,t) = x_k$ for all $t \geq s$ and $\zeta_{i,j}(s,s) = x_i$.

Using (6.6.8), one may obtain $\mu_{i,u(j)}(s,t)$ by taking derivatives with respect to x_u and then setting $x_v = 1$ for $v = i,\ldots,k$. Note from (6.6.8) that $\zeta_{i,j}(s,t) = \zeta_{i,j}(t-s)$ so that $\mu_{i,u(j)}(s,t) = \mu_{i,u(j)}(t-s)$ depends on s and t through $t - s$.

Let $\epsilon_{ij} = b_{ij} - d_{ij}, i = 1,\ldots,k-1, j = 1,\ldots,m$. Taking derivatives of both sides of (6.6.8), putting $x_v = 1, v = i,\ldots,k$ and noting that $\zeta_{i,j}(s,t) = 1$ when $x_v = 1, v = i,\ldots,k$, one obtains for $t_{j-1} \leq s < t_j \leq t, j = 1,\ldots,m$ and $i \leq u, i, j = 1,\ldots,k-1$:

$$\frac{d}{dt}\mu_{i,u(j)}(s,t) = \epsilon_{ij}\mu_{i,u(j)}(s,t) + \alpha_{i+1j}\mu_{i+1,u(j)}(s,t), \tag{6.6.9}$$

where $\mu_{i,u(j)}(s,s) = \delta_{iu}$ (the Kronecker's δ) and $\mu_{i+1,i(j)}(s,t)$ is defined as $\mu_{i+1,i(j)}(s,t) = 0$.

From (6.6.9), for $t_{j-1} \leq s < t_j \leq t, j = 1,\ldots,m$ and $i = 1,\ldots,k-1$,

$$\mu_{i,i(j)}(s,t) = \mu_{i,i(j)}(t-s) = exp[\epsilon_{ij}(t-s)], \tag{6.6.10}$$

and for $1 \leq i < u \leq k-1$ and $t_{j-1} \leq s < t_j \leq t, j = 1,\ldots,m$,

$$\mu_{i,u(j)}(s,t) = \mu_{i,u(j)}(t-s) = \alpha_{i+1j}\int_s^t \mu_{i+1,u(j)}(x-s)exp[\epsilon_{ij}(t-x)]dx$$
$$= \alpha_{i+1j}\int_0^{t-s} exp(\epsilon_{ij}y)\mu_{i+1,u(j)}(t-s-y)dy. \tag{6.6.11}$$

From (6.6.10) and (6.6.11), we have that for $t_{j-1} \leq s < t_j \leq t, j = 1,\ldots,m$ and for $i < u, i, u = 1,\ldots,k-1$,

$$\mu_{i,u(j)}(s,t) = [\prod_{v=1}^{u-i}\alpha_{i+vj}]\int_{y_0}^{t-s}\int_{y_1}^{t-s}\cdots\int_{y_{u-i-1}}^{t-s} exp[(t-s-y_{u-i})\epsilon_{uj}]$$

$$\{\prod_{v=1}^{u-i}exp[(y_v - y_{v-1})\epsilon_{i+v-1j}]dy_v\}, \tag{6.6.12}$$

where $y_0 = 0$.

On substituting (6.6.12) into (6.6.7), one may compute $m_{k-1}(x,t)$. Then, using (6.6.5), $\lambda(t)$ can readily be approximated.

6.7 SOME POSSIBLE APPLICATIONS OF MULTIEVENT MODELS

As illustrated in Section (6.1), much biological evidence suggests that some animal and human cancers would be better described by multi-event models than by two-stage models. Since different models would in general give very different results in assessing effects of environmental agents (26, 27), in cases where multievent models provide better fitting, multievent models should be adopted for risk assessment of environmental agents. This is especially important if the environmental agents affect the mutation rates and/or cell proliferation rates of intermediate cells. Since the multievent models were only recently proposed, no attempts have been made in this area of research. Definitely many new studies are required to develop conditions under which the multievent models would provide significantly better risk assessment results than the MVK two-stage models.

Alternatively, since the multievent models include more than two stages which allow for the possibility of entertaining various models, the multievent model is also useful for discriminating between different models. In fact, Chu et al. (8) have used the multievent models to discriminate between various models involving cancer promotion; for more detail, the readers are referred to Chu et al. (8).

6.8 CONCLUSIONS AND SUMMARY

Much biological data suggest that in some cancers the process of carcinogenesis involves more than two stages. This is true especially of cancers which involve both oncogenes and antioncogenes. Specific examples include human colon cancer which involves at least two oncogenes (myc and ras) and at least three antioncogenes (see 28), and human lung cancer which involves at least five oncogenes (myc, H-ras, K-ras, raf, jun) and at least four antioncogenes (see 30). Because of these results, in this chapter we thus proceed to develop the mathematical theories of multievent models and indicate some of their applications.

To start the chapter, in Section (6.1) we give some biological evidence supporting the multievent models which may involve more than two stages. As illustrated in Section (6.1), these models arise because in some cases immortalization and transformation are not sufficient to induce tumorigenic conversion of normal stem cells under usual conditions; at least one more event is required. This is especially true of cancers which involve both oncogenes and antioncogenes.

Because of the biological evidence, in Section (6.2) multievent models which may involve more than two stages are proposed and the probability generating functions (PGF) of the number of tumors for these models are derived under some general conditions. In Section (6.2) it is shown that the multievent model of carcinogenesis is in fact a filtered Poisson process under some general conditions. By using the PGFs of the number of tumors in Section (6.2), we derive in Section (6.3) the age-specific incidence functions of tumors and give some approximations to these incidence functions by assuming that the mutation rates are very small and that the cell proliferation rates of intermediate cells are small (≤ 0.1). By using the PGFs of the number of tumors in Section (6.2), the expected numbers of tumors and the variances and covariances of the numbers of tumors are derived in Section (6.4) by solving some ordinary differential equations.

Having derived the mathematical theories for general multievent models, in Sections (6.5) and (6.6) we then consider two special cases of multievent models in each of which the incidence functions are derived. The special cases of multievent models include the homogeneous multievent models and the nonhomogeneous piece-wise multievent models which are analogs of the Tan-Gastardo two-stage models. The classical Armitage–Doll multistage model is a special case of the homogeneous multievent model if one ignores cell proliferation of intermediate cells. Finally some possible applications of multievent models are indicated in Section (6.7).

REFERENCES

1. Armitage, P. and Doll, R., The age distribution of cancer and multistage theory of carcinogenesis, British Jour. of Cancer **8** (1954), 1-12.

2. Armitage, P. and Doll, R., Stochastic models for carcinogenesis., Univ. California Press, Berkeley, Cal., In: "Fourth Berkeley Symposium on Mathematical Statistics and Probability" (1961), 19-38.
3. Barrett, J.C. et al., Oncogene and chemical induced neoplastic progression: Role of tumor suppression., National Inst. Environmental Health Science, NIH, Research Triangle Park, N.C. (1987).
4. Bodmer, W.F. et al., Localization of the gene for familial adenomatous polyposis on chromosome 5, Nature **328** (1987), 614-616..
5. Bos, J.L. et al., Prevalence of ras gene mutations in human colorectal cancers., Nature **327** (1987), 293-297..
6. Compere, S. J. et al., The ras and myc oncogenes cooperate in tumor induction in many tissues when introduced into midgestation mouse embryos by retroviral vectors, Proc. Natl. Acad. Sci. USA **86** (1989), 2224–2228.
7. Chu, K.C., Multi-event model for carcinogenesis: A model for cancer causation and prevention., Mass MJ, Kaufman DG, Siegfied JM, Steele VE, Nesnow S (Eds), Raven Press New York, In:"Carcinogenesis: A Comprehensive Survey Volume 8: Cancer of the Respiratory Tract-Predisposing Factors" (1985), 411-421.
8. Chu, K.C., Brown, C.C., Tarone, R. and Tan, W.Y., Differentiating between proposed mechanisms for tumor promotion in mouse skin using the multi-event model for cancer, J. Nat. Cancer Inst. **79** (1987), 789-796.
9. Forrester, K. et al., Detection of high incidence of k-ras oncogenes during human colon tumorigenesis., Nature **327** (1987), 298-303.
10. Franza, B. R. et al., In vitro establishment is not a sufficient prerequisite for transformation by activated ras oncogenes, Cell **44** (1986), 409–418.
11. Land, H., Parada, L. F. and Weinberg, R. A., Tumorigenic conversion of primary embryofibroblasts requires at least two cooperating oncogenes, Nature **304** (1983), 596–601.
12. Moolgavkar, S.H., Carcinogenesis modeling: From molecular biology to epidemiology, Ann. Rev. Public Health **7** (1986), 151-169.
13. Moolgavkar, S.H. and Knudson, A.G., Mutation and cancer: A model for human carcinogenesis, J. Nat. Cancer Inst. **66** (1981), 1037-1052.
14. Oshimura, M., Gilmer, T.M. and Barrett, J.C., Nonrandom loss of chromosome 15 in Syrian hamster tumors induced by v-Ha-ras plus v-myc oncogenes, Nature **316** (1985), 636-639.
15. Ruley, H.E. et al., Multistep transformation of an established cell line by the adenovirus Ela and T24 Ha-ras-1 genes, "In "Cancer Cells

3, Growth Factors And Transformation,"," Edited by Feramisco, J. Ozannr, B. and Stiles, C., Cold Spring Harbor Lab, 1985.

16. Sinn, E. et al., Coexpression of MMTV/v-Ha-ras and MMTV/c-myc genes in transgenic mice: Synergistic action of oncogenes in vivo, Cell **49** (1987), 465–475.

17. Solomon, E. et al., Chromosome 5 allele loss in human colorectal carcinomas, Nature **328** (1987), 616-619.

18. Tan, W.Y., A stochastic Gompertz birth-death process, Statist. and Prob. Lett. **4** (1986), 25-28.

19. Tan, W. Y. and Gastardo, M. T. C., On the assessment of effects of environmental agents on cancer tumor development by a two-stage model of carcinogenesis, Math. Biosciences **73** (1985), 143–155.

20. Tan, W. Y. and Brown, C. C., A nonhomogeneous two-stage model of carcinogenesis, Math. Modelling **9** (1987), 631–642.

21. Tan, W. Y., Chu, K. C. and Brown, C. C., Multievent model with stochastic birth and death, National Cancer Inst./NIH, DCPC/BB report, Bethesda, MD.

22. Tan, W.Y. and Piantadosi, S., On stochastic growth models with application to stochastic logistic growth, To appear in Statistica Sinica, 1991.

23. Tan, W.Y. and Singh, K., A mixed model of carcinogenesis-with applications to retinoblastoma, Math. Bioscience **98** (1990), 201–211.

24. Thomassen, D.G. et al., Evidence for multiple steps in neoplastic transformation of normal and preneoplastic Syrian hamster embryo cells following transfection with Harvey murine sarcoma virus oncogene (v-Ha-ras), Cancer Research **45** (1985), 726-732.

25. Tsunokawa, Y. et al., Integration of v-ras does not necessarily transform an immortalized murine cell line, Gann **75** (1984), 732-736.

26. Van Ryzin, J., Quantitative risk assessment, Jour. Occup. Medicine **22** (1980), 321-326.

27. Van Ryzin, J., Low dose risk assessment for regulating carcinogens, Center for the Social Sciences Newsletter, Columbia University, N.Y. **4, No. 1** (1984), 1-5.

28. Vogelstein, B., et al., Genetic alterations during colorectal-tumor development,, The New England Jour. of Medicine **319** (1988), 525–532.

29. Weinberg, R. A., Oncogenes, antioncogenes and the molecular bases of multistep carcinogenesis, Cancer Research **49** (1989), 3713–3721.

30. Weston, A. et al., Differential DNA sequence deletions from chromosomes 3, 11, 13 and 17 in squamous-cell carcinoma, large-cell carcinoma, and adenocarcinoma of the human lung., Proc. Natl. Acad. Sci. USA **86** (1989), 5099– 5103.

Index